Membrane Technology in Water Treatment in the Mediterranean Region (ProMembrane)

Membrane Technology in Water Treatment in the Mediterranean Region (ProMembrane)

Antonia Lorenzo and Aníbal Vega

IWA Publishing
London • New York

Pro Membrane

Published by **IWA Publishing**
Alliance House
12 Caxton Street
London SW1H 0QS, UK
Telephone: +44 (0)20 7654 5500
Fax: +44 (0)20 7654 5555
Email: publications@iwap.co.uk
Web: www.iwapublishing.com

First published 2011

Typeset in India by MPS Limited, a Macmillan Company, Chennai, India, www.macmillansolutions.com

British Library Cataloguing in Publication Data
A CIP catalogue record for this book is available from the British Library

Library of Congress Cataloging-in-Publication Data
A catalog record for this book is available from the Library of Congress

ISBN13: 9781843393702
ISBN: 1843393700

Contents

Preface

Gerhard Schories

Water is the most essential resource of life. About 97,5 % of the World's water resources (1.386 Mio km^3) are salty seawater, not usable for agriculture, technical purposes or simply quenching the thirst of about 6 Billion people on Earth. Just 2.5% of the world's water is fresh water, but only 30 % of it are accessible for us, either as groundwater or as surface water bodies. These fresh water sources are not equally distributed, there are regions in the world with plenty of good water and other, often highly populated regions, where fresh water is a very scarce good.

Middle Eastern and Northern African (MENA) countries belong to those of the world. A growing population in combination with demand for irrigation in agriculture leads to a dilemma, that will seriously affect food and water supply security of more than 340 Million people in that region. Global warming and resulting climate change is even worsening the situation, accelerated by an increasing population (in 2025 about 430 Million people are expected to live in the target region). But there will be no increase in the amount of available water unless we all work on that issue.

In order to solve the problems of water scarcity a threefold approach is useful:

- Appropriate technology
- Exchange of know-how and information
- Applied research and development

Membrane technology is a powerful technical tool to significantly improve the water supply situation in the MENA region. Membranes can be used for producing potable water from seawater. They can also be used for purifying wastewater for reuse, either in a combination of different membrane separation processes, or in combination with other technologies, such as biological treatment. PROMEMBRANE was initiated to provide a platform to collect and exchange information. Furthermore, it was planned to create a pool of experts which could be contacted in case of special expertise is required. As a technology transfer institution it is the aim of ttz-Bremerhaven to promote and support such activities.

In addition, it was our intention to motivate young scientists in working on application of membrane technology and presenting actual work results in the frame of a Young Scientist Award. As the coordinator, but as well as member of the evaluation committee it was a great pleasure to see work of young researcher with high relevance, done with great commitment and engagement and showing useful results. I am quite convinced that with such commitment we can successfully meet the challenge. I also hope, that this publication summarizing the main activities and outcomes of PROMEMBRANE as well as the project website www.promembrane.info provide useful source of information to initiate and to carry out future research to develop sustainable solutions to solve the water scarcity problem in the MENA countries.

Dr.-Ing. Gerhard Schories
Technical Director
Water, Energy and Landscape Management
ttz Bremerhaven, Germany
Coordinator of ProMembrane

Contributors

Antonia Lorenzo
Bioazul S.L.
Severo Ochoa 7
29590 Malaga (Spain)

Aníbal Vega
Bioazul S.L.
Severo Ochoa 7
29590 Malaga (Spain)

Dr.-Ing. Gerhard Schories
Technical Director Institute of Water
Energy and Landscape Management
ttz Bremerhaven
An der Karlstadt 6
27568 Bremerhaven
Germany
and International Consultant to Prince Khalid Bin Sultan Chair for Water Reuse at King Saud University, Riyadh, KSAW

Julie Hernandez
Institute of Water, Energy and Landscape Management
ttz Bremerhaven
An der Karlstadt 6
27568 Bremerhaven
Germany

Dr. Alessandra Criscuoli
Institute on Membrane Technology (ITM-CNR)
c/o University of Calabria
via P. Bucci cubo 17/C
Rende (CS) 87030
Italy

Dr. Alberto Figoli
Institute on Membrane Technology (ITM-CNR)
c/o University of Calabria
via P. Bucci cubo 17/C,
Rende (CS) 87030
Italy

Prof. Enrico Drioli
Institute on Membrane Technology (ITM-CNR)
c/o University of Calabria
via P. Bucci cubo 17/C
Rende (CS) 87030
Italy
and Department of Chemical Engineering and Materials
via P. Bucci cubo 44/A
Rende (CS) 87030
Italy

Dr.-Eng Rashed Al-Sa`ed
Associate Professor Sanitary and Environmental Engineering
Institute of Environmental and Water Studies
Birzeit University
P.O. Box 14
Birzeit, West Bank
Palestine

Dr. Ziad Mimi
Associate Professor and Director
Institute of Environmental and Water Studies
Birzeit University
P.O. Box 14
Birzeit, West Bank
Palestine

Prof. Sami Sayadi
Head Laboratory Bioprocesses
Centre of Biotechnology at Sfax
Route Sidi mansour Km 6
P.O. Box 1177
3018 Sfax
Tunisia

Prof. Dr. Ing. Adnan J. Ghata
Department of Chemical Engineering
Al-Baath University
P.O. Box 77
Homs
Syrian Arab Republic

Prof. Dr. Ing. Fouad Atallah
Department of Chemical Engineering
Al-Baath University
P.O. Box 77 Homs
Syrian Arab Republic

Prof. Dr.Hussein Abdel-Shafy
Water Research & Pollution Control Department National Research Centre
El Buhouth St.
Dokki
12311 Cairo
Egypt

Introduction

Antonia Lorenzo and Aníbal Vega

WATER SCARCITY IN THE MEDITERRANEAN REGION

The current water crisis in the Mediterranean region is continuously escalating. While representing 5% of the total world population, the Middle East & North Africa (MENA) region contains only 0.9% of global water resources (The World Bank, 1996). As it can be observed in Figure 1, the actual renewable fresh water resources are in the Middle East and North Africa region the lowest in the world (The World Bank, 2007). The number of water-scarce countries in the MENA region has increased from 3 in 1955 (Bahrain, Jordan and Kuwait) to 11 by 1990 (with the inclusion of Algeria, Israel, Palestine, Qatar, Saudi Arabia, Somalia, Tunisia, the United Arab Emirates and Yemen). Another 7 are anticipated to join the list by 2025 (Egypt, Ethiopia, Iran, Libya, Morocco, Oman and Syria) (Darwish, 1994). In particular in the Middle East region, most of the water that runs is located in three major waterways: the Tigris-Euphrates, Nile and Jordan River systems. Mutual dependence on these resources has made water a catalyst for conflict, occurring 21 armed disputes over water in recent history (i.e. 1967 War, the Iran-Iraq War, Israel-Palestinian Authority) (Darwish, 1994) has been recognised the important role that water plays in order to foment interstate relations and therefore numerous attempts of resolution have taken place, including diplomatic, bilateral and multilateral treaty efforts, ranging from the 1959 Agreement for the Full Treaty Utilisation of Nile Waters to the 1994 Israeli-Jordanian (Berman *et al.*, 1999).

Despite these agreements, nations have begun to come into conflict over water promoted by a variety of factors: (1) with population rates among the highest in the world, countries in the MENA region are consuming water at a much higher rate than can be replenished naturally; (2) in an area already critically short of water, this depletion has been compounded by domestic pollution, which has contributed to a deterioration of usable resources and a general decline in the quality of available water; and (3) expanding initiatives in agriculture and industry have further eroded regional water availability (Berman *et al.*, 1999).

The average water availability per capita in the region is about 1,200 cubic meters per year (world average is close to 7,000). The annual water availability in the region ranges from a high of about 1,800 cubic meters per person in Iran to less than 200 cubic meters per person in Jordan, Palestine, and Yemen. By 2025, the regional average water availability is projected to be just over 500 cubic meters/person/year (Rendel *et al.*, 2006). As shown in Figure 2, already in 2005 in the Mediterranean Region almost all countries from North Africa were suffering from water scarcity, while others in the east region were approaching to it and the European ones were suffering for little water scarcity (International Water Management Institute, 2008).

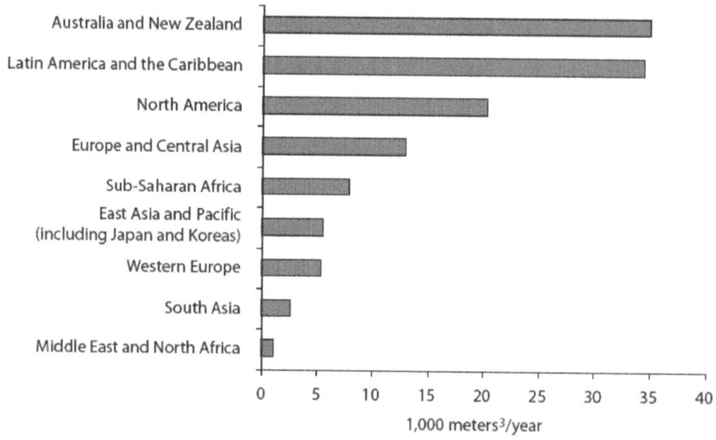

Figure 1 Actual renewable freshwater resources per capita, by region (Source: MENA Development Report, The World Bank, 2007)

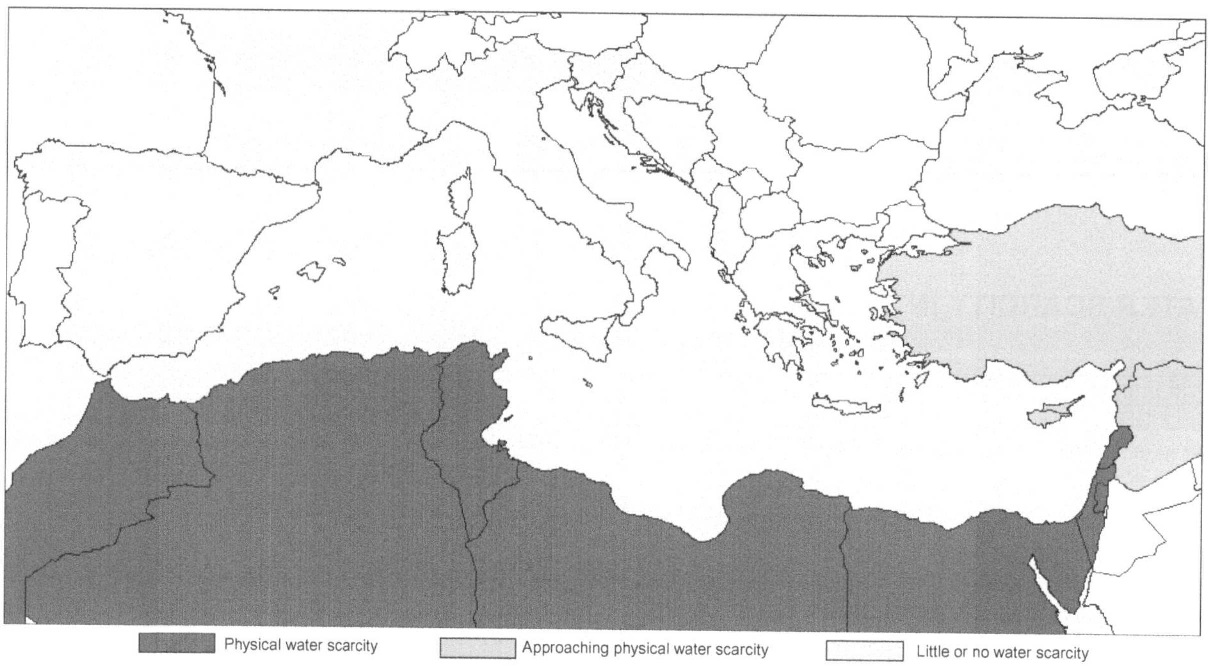

Figure 2 Water scarcity in the Mediterranean region (based on IWMI Annual Report, 2006/2007)

In spite water shortage, as access to a safe water supply and adequate sanitation are both essential to a healthy life, most North African and Eastern Mediterranean countries had made a major effort to improve drinking water supply and sanitation services. However, there are still major deficits in rural areas in countries such as Morocco, Tunisia and Syria where the access to drinking water is less than 65% mainly because their financial situation hinders from establishing pipes lines to small villages and remote areas.

Past surveys of the technical options for sustainable water supply in the Mediterranean Partner Countries (MPCs) shows that there are well proven technologies available which could foster an efficiency revolution in all sectors of the economy (i.e. agriculture, industry, services sectors). For instance, wastewater reuse and reclamation can constitute an additional water resource for irrigation, as well as desalination of sea water to provide potable water. The application of advanced water treatment methods, as membrane technology, is the key factor to abate the water escalating crisis existing today.

This limiting Mediterranean particular feature, the scarcity of water resources, poses a barrier for the sustainable development of this region, as well as for the increment of the quality of life and peace. The water situation of the southern and eastern countries of the Mediterranean Sea is dramatically serious, which shows rapidly increasing population, increasing irrigated areas and low rainfall. A number of areas of the north will also face water scarcity, such as large urban areas and coastal and tourism settlements, particularly eastern Spain, South Italy, Greek Islands and

Malta. The non-renewable water reservoirs are diminishing rapidly; the remaining resources are threatened by seawater intrusion and by pollution from contaminated effluents. The dramatic water supply situation in the target regions are highlighted by the fact that in some areas of Greece, Tunisia and Palestine less than 20% of the remaining water resources have drinking water quality. ProMembrane has served as a **tool for dissemination of research and development activities** of one of the technologies, which due to its technical advantages, represent the key for water treatment and reuse: **membrane technology** (MT).

Increased population growth, rapid urbanization and industrialization associated with living standards improvement in most MENA countries have aggravated the water balance gap between the available water supplies and the water demands. In MENA countries, especially those with limited water resources, there are challenges of satisfying rapid and substantial increases in water demand for industrial, domestic and agricultural purposes. The annual precipitation in MENA countries ranged between 150 and 600mm, while the available water resources are limited, overexploited, polluted and/or politically confronted. Recent survey on the present water uses in these regions revealed an average of 22% for municipal use, 3% for industrial use and 75% for agriculture use. Non-conventional resources such as seawater and brackish water desalination, in addition to limited reclaimed effluent can alleviate the looming water crises in the MENA area. In the Gulf region and South Mediterranean countries, experts have already realized the nature of these challenges and recommended several key policies for water stressed countries, of which are reduction of water subsidies, the increment of reclaimed effluent reuse, aquifer recharge, advancement of local industrial bases and building capacity in sustainable membrane processes for both water and wastewater treatment (Al-Sa`ed *et al.*, 2009).

Membrane technology can be a very powerful and flexible tool for developing technological concepts to purify and recycle polluted water streams or to produce potable water from seawater.

WASTEWATER TREATMENT USING MEMBRANES

Wastewater treatment by membrane operations is an interesting way to recover purified water from the effluent streams coming from industries, by improving the overall performance of the production process. This is the case, for example, of textile industries that discharge aqueous solutions containing dyes to the environment. By treating (in fact there is no 'treatment', but a separation) these streams by ultrafiltration, nanofiltration, reverse osmosis, membrane distillation, it is possible to obtain a water stream without dyes and a stream concentrated in dyes that can be recycled to the process (Calabrò *et al.*, 1990; Ciardelli *et al.*, 2000; Marcucci *et al.*, 2002; Tang and Chen, 2002; Purkait *et al.*, 2004; Mozia *et al.*, 2005; Mozia *et al.*, 2007; Van der Bruggen *et al.*, 2004; Banat *et al.*, 2005; Criscuoli *et al.*, 2008). Other interesting cases are the recovery of heavy metals from water by extraction by membrane contactors or by using supported liquid membranes (Yun *et al.*, 1993; Ho *et al.*, 2000; Ho and Poddar, 2000; Lin and Juang, 2001), the volatile organic compounds (VOCs) removal by pervaporation (Peng *et al.*, 2003) and the arsenic removal by nanofiltration or reverse osmosis (Brandhuber and Amy, 1998; Urase *et al.*, 1998; Kang *et al.*, 2000; Ning, 2002; Vrijenhoek and Waypa, 2000; Saitua *et al.*, 2005).

MEMBRANE BIOREACTOR TECHNOLOGY

The technology of membrane separation of activated sludge, commonly referred to as 'membrane bioreactor' (MBR), is considered by the European Union as one of the Best Available Technologies (BAT) for municipal and industrial wastewater treatment and the increased interest in the MBR technology is clearly evident from the high numbers of projects of research funded (European IPPC Bureau).

MBRs combine membrane filtration with biological treatment and are effective for treating industrial wastewater, domestic wastewater and specific municipal wastewater, where a small footprint, water reuse, or stringent discharge standards are required. Promising applications also exist in treating landfill leachate, chlorinated solvents in manufacturing wastewaters, wastewater polluted with trace pharmaceutical compounds, for groundwater remediation, and for the treatment of wastewaters from petrochemical industry. Compactness, reduced sludge production and high efficiency in physical disinfection are the main advantages claimed for these systems (Qin *et al.*, 2007). Since the early MBR installations in the 1990s, the number of MBR systems has grown considerably. Total European revenue for the MBR market was around 40 million Euro in 2005 with a steady growth rate of 9%. The European Commission decided in 2005 to boost the MBR technology in Europe through the 6th Framework Program while financing 4 projects entirely dedicated to the development of the MBR technology for municipal applications. These 4 projects (namely AMEDEUS, EUROMBRA, MBR-TRAIN and PURATREAT) were performed over the period 2005–2009 and formed the projects coalition 'MBR-Network' (www.mbrnetwork.eu). € 15 millions were invested over 4 years in this R&D initiative gathering about 50 European and international companies and institutions (Lesjean *et al.*, 2006). The MBR market growth is shown in Figure 3. A detailed review on MBRs is available at the website of the European project MEDINA (http://medina.unical.it).

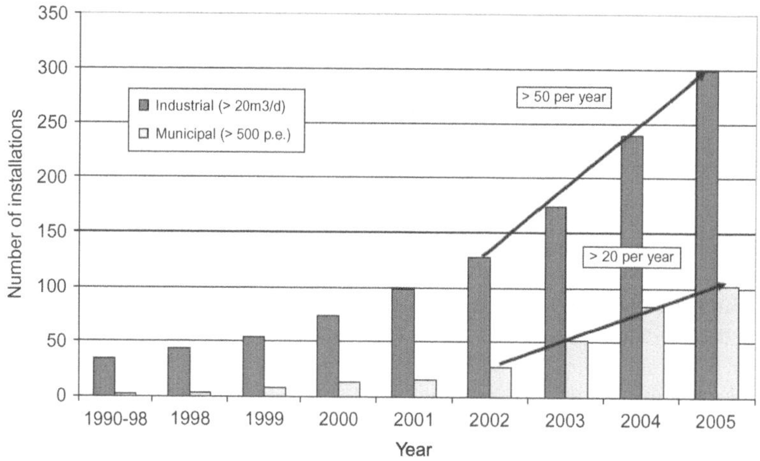

Figure 3 Development of industrial and municipal MBR markets (402 references in graph) (Lesjean *et al.*, 2007)

Due to the cost reduction and the increased confidence, a relative greater number of larger plants should be seen in the coming years (Lesjean *et al.*, 2007).

The installed membrane surface confirms that although ¾ of the plants were erected for industrial schemes, given the larger capacity of the municipal plants, the revenue of the European MBR market originated mostly from the municipal sector since 2002. In the period 2003–2005, about 400,000m^2 of membranes were installed in municipal applications, and only 65,000m^2 in industrial plants.

The industrial market is highly competitive over the entire range of plant capacity and wastewater types. The main drivers has remained essentially the national and European regulations (Integrated Pollution Prevention Control Directive 96/61/EEC, Urban Waste Water Treatment Directive 91/271/EEC), and the economical pressure towards internal water recycling (Lesjean *et al.*, 2007). But this market is still growing and not mature yet. An increased number and capacity of plants is expected, as well as an increased competition. The main drivers for further growth will be: the European Water Framework Directive (WFD), Urban Waste Water Treatment Directive (91/271/EEC), and Bathing Water Directive, but also increased competitiveness (technological improvement, standardization etc), lack of place, retrofitting of existing plants and water scarcity (water reuse or recycling) (Lesjean *et al.*, 2007).

The already mentioned web platform of the coalition 'MBR-Network' is designed to complement this offer while providing actual on-line information on MBR technology. In particular, it includes an extensive database of articles and reports, as well as a discussion forum accessible by the international MBR community (about 400 members were registered over the first 9 months of existence). Such public communication platforms and networks are to optimize the diverse research activities; accelerate the set up of good practices and increase awareness and acceptance towards the technology. This will positively accompany the global growth of the MBR market.

SEAWATER DESALINATION USING MEMBRANES

An intensive application of membrane operations is also in the water desalination field. Today, reverse osmosis (RO) is the dominant technology and all new installed plants use RO units for desalinating water. Moreover, as a pre-treatment step, microfiltration and ultrafiltration units are more often present, instead of more traditional coagulation or sand filtration devices (Glueckstern *et al.*, 2001).

In the Middle-East desalination capacity is rapidly expanded as water demands soar. Saudi Arabia produces 3 million m^3 a day, which is expected to jump to more that 9 million in 20 years. Similarly, Israel is planning to nearly double its production capabilities by 2020 from 400 million m^3 to more than 750m^3 (World Wildlife Fund, 2007). The number of membrane-based desalination plants is, therefore, increasing worldwide, accounting for about 80% of all desalination facilities (Lesjean *et al.*, 2007).

The investments in RO technology are increasing, as well as the number of projects on seawater desalination. Global desalination capacity is expected to reach 107 million m^3 per day in 2016 and it has been calculated that over the decade 2007–2016 $64 billion will be spent on building desalination plants (IDA Water News, 2008). This scenario states the extreme importance for Mediterranean countries to 'take part' to the evolutions in the field and, therefore, to overcome the existing gap with more-advanced countries.

Larger plants were built and also bigger RO modules were designed and developed (16 inch modules with membrane areas of 1570–1725ft^2 instead of the traditional 8 inch modules with membrane areas of 365–400ft^2), both to reduce operating and capital costs (because of the capability to operate at higher flow rates) and to work with higher capacities (IDA Water News, 2008). Some of the world's largest sea water reverse osmosis (SWRO) plants are reported in Table 1.

Table 1 Some of the world's largest SWRO plants

SWRO plant	Capacity (MGPD)
Jubail, United Arab Emirates	20
Madina-Yanbu, United Arab Emirates	33.3
Fujairah, United Arab Emirates	45
Ashkelon, Israel	84.4
Tampa Bay, United States	25
Trinidad and Tobago	28.8
Carboneras, Spain	32
Rishon Letzion, Israel	100
Singapore	36

On January 2008, Spain's Environment Minister inaugurated the Europe's largest RO desalination plant at Valdelentisco in Murcia (Industry News, 2008), and another large RO plant is taking shape in Barcelona (IDA Water News, 2008).

REFERENCES

Al-Sa`ed R., Sayadi S., Ghata A., Abdel-Shafy H., Schories G., Oropeza M., Lorenzo A. and Drioli E. (2009). Advancing membrane technologies for wastewater treatment and reclamation in selected Arab MENA countries. *Desalin. Water Treat.*, **4**, 287–293.

Banat F., Al-Asheh S. and Qtaishat M. (2005). Treatment of waters colored with methylene blue dye by vacuum membrane distillation. *Desalination*, **174**, 87–96.

Berman I. and Wihbey P. (1999). The new water politics of the Middle East. *Strategic Review*. The Institute for Advanced Strategic & Political Studies. http://www.iasps.org/strategic/water.htm (accessed 22 February 2010).

Brandhuber P. and Amy G. (1998). Alternative methods for membrane filtration of arsenic from drinking water. *Desalination*, **117**, 1–10.

Calabrò V., Pantano G., Kang M., Molinari R. and Drioli E. (1990). Experimental study on integrated membrane processes in the treatment of textile effluents – energy and exergy analysis. *Desalination*, **78**, 257–277.

Ciardelli G., Corsi L. and Marcucci M. (2000). Membrane separation for wastewater reuse in the textile industry. *Resources Conserving and Recycling*, **31**, 189–197.

Criscuoli A., Zhong J., Figoli A., Carnevale M. C., Huang R. and Drioli E. (2008). Treatment of dye solutions by vacuum membrane distillation. *Water Res.*, **42**, 5031–5037.

Darwish A. (1994). *Water Wars*. Lecture given at the Geneva Conference on Environment and Quality of Life, Geneva, Switzerland, June 2, 1994.

Glueckstern P., Thoma A. and Prielm M. (2001). The impact of R&D on new technologies, novel design concepts and advanced operating procedures on the cost of water desalination. *Desalination*, **139** (1–3), 217–228.

Ho W. S. W., Kalini S. and Poddar T. K. (2000) *Proceedings of the AIChE Spring National Meeting*, 5–9 March 2000, Atlanta, GA, USA, pp 10–15.

Ho W. S. W. and Poddar T. K. (2000). *Proceedings of the AIChE Spring National Meeting*, Atlanta, GA, USA, pp. 38–43.

Industry News (2008). *IDA Water News* 3.

International Water Management Institute (2008). Coping with Competition for Water: Improving Basin Water Management and Water Productivity. International Water Management Institute Annual Report 2006/2007. http://www.iwmi.cgiar.org/About_IWMI/Strategic_Documents/Annual_Reports/2006_2007/theme1.html (accessed 15 February 2010).

Kang M., Kawasaki M., Tamada S., Kamei T. and Magara Y. (2000). Effect of pH on the removal of arsenic and antimony using reverse osmosis membranes. *Desalination*, **131**, 293–298.

Lesjean B. and Huisjes E. H. (2007). Survey of European MBR market, trends and perspectives. In: *IWA 4th International Membrane Technologies Conference*, 15–17 May 2007, Harrogate, UK.

Lesjean B. and Luc F. (2006) Assessment of the membrane bioreactor technology and European market outlook. http://www.idswater.com/Common/Paper/Paper_155/ (accessed 12 May 2007).

Lin S.-H. and Juang R. S. (2001). Mass-transfer in hollow-fiber modules for extraction and back-extraction of copper(II) with LIX64N carriers. *J Membr. Sci.*, **188**, 251–262.

Marcucci M., Ciardelli G., Matteucci A., Ranieri L. and Russo M. (2002). Experimental campaigns on textile wastewater for reuse by means of different membrane processes. *Desalination*, **149**, 137–143.

MEDINA: Membrane-based desalination: an integrated approach. A research project supported by the European Commission under the Sixth Framework Programme. Project reference 36997. http://medina.unical.it (accessed 7 February 2010).

Membrane Bioreactor Technology Network. www.mbrnetwork.eu (accessed several times in February and March 2010).

Mozia S., Tomaszewska M. and Morawski A. W. (2005). A new photocatalytic membrane reactor (PMR) for removal of azo-dye Acid Red 18 from water. *Appl. Catal. B: Environ.*, **59**, 133–139.

Mozia S., Tomaszewska M. and Morawski A. W. (2007). Photodegradation of azo dye Acid Red 18 in a quartz labyrinth flow reactor with immobilized TiO_2 bed. *Dyes Pigment.*, **75**, 60–66.

Ning Y. R. (2002). Arsenic removal by reverse osmosis. *Desalination*, 143, 237–241.

Peng M., Vane L. M., and Liu S. X. (2003). Recent advances in VOCs removal from water by pervaporation. *J. Hazard. Mater.*, **98**, 69–90.

Purkait M. K., DasGupta S. and De S. (2004). Removal of dye from wastewater using micellar-enhanced ultrafiltration and recovery of surfactant. *Sep. Purif. Technol.*, **37**, 81–92.

Qin J.-J., M. H. Oo, Tao G. and Kekre K. A. (2007). Study on petrochemical wastewater treatment and reuse using submerged MBR. *J. Membr. Sci.*, **293**, 161–166.

Renger J. and Rothenberger D. (2006). Investment needs and role of the private sector. *Paper presented at the Regional Water Forum Water Governance in the MENA Region – Current Situation*, July, Yemen.

Saitua H., Campderros M., Cerutti S. and Padilla A. P. (2005). Effect of operating conditions in removal of arsenic from water by nanofiltration membrane. *Desalination*, **172**, 173–180.

Tang C. and Chen V. (2002). Nanofiltration of textile wastewater for water reuse. *Desalination*, **143**, 11–20.

The European IPPC Bureau. http://eippcb.jrc.es/ (accessed 15 February 2010)

The World Bank Report (1996). *From Scarcity to Security: Averting a Water Crisis in the Middle East and North Africa*. World Bank Publications, Washington, DC.

The World Bank, International Bank for Reconstruction and Development (2007). *Making the Most of Scarcity Accountability for Better Water Management Results in the Middle East and North Africa*. MENA Development Report. The World Bank, Washington DC, chapter 1, pp. 1–31.

Urase T., OH J. and Yamamoto K. (1998). Effect of pH on rejection of different species of arsenic by nanofiltration. *Desalination* **117**, 11–18.

Van der Bruggen B., Curcio E. and Drioli E. (2004). Process intensification in the textile industry: the role of membrane technology. *J. Environ. Manage.*, **73**, 267–284.

Vrijenhoek E. M. and Waypa J. (2000). Arsenic removal from drinking water by a "loose" nanofiltration membrane. *Desalination*, **130**, 265–277.

Yun C. H., Prasad R., Guha A. K. and Sirkar K. K. (1993). Hollow fiber solvent extraction removal of toxic heavy metals from aqueous waste streams. *Ind. Eng. Chem. Res.*, **32**, 1186–1195.

Chapter 1

The ProMembrane project

Antonia Lorenzo and Aníbal Vega

1.1 DESCRIPTION OF THE PROJECT

ProMembrane – Promotion and focussing of current research activities of Membrane Technology in Water Treatment in the Mediterranean Region – was funded by the 6th Framework Programme of the European Union under its priority B1.3 that addressed Specific Support Actions for Mediterranean Partner Countries. The project was developed during two years in a coordinated consortium of seven specialized-in-water treatment partners from Germany, Tunisia, Italy, Spain, Egypt, Syria and Palestine to be successfully completed in August 2008.

As a Specific Support Action (SSA), the primary objective of ProMembrane was **to support and focus the current research and development activities in membrane technology for water treatment** in the Mediterranean area, promoting international cooperation among research organizations and universities devoted to membrane technologies. It aimed the overture and presentation of the on-going research to facilitate the dissemination and to make known the existing potential for the development of this technology in the foreseen regions. The project was carried out by seven research institutions with the overall goal of encouraging further research in the area, increase the scientific background and unifying efforts in the membrane research field in order to offer in a medium-long term an option to the water supply and treatment situation in the Mediterranean water scarce area.

The **general objectives** addressed in ProMembrane were:

- To support research activities in membrane technology for water treatment in the Mediterranean region.
- To promote and facilitate the dissemination, transfer and exploitation of results of current and past research, by the organisation of regional fairs and a final conference.
- To encourage the permanent participation of young scientists in the field of membrane technology.
- To promote the establishment of permanent communication links between research centres in the Mediterranean region, promoting the know-how exchange and expertise in relation to membrane technologies.

The **specific scientific and technologic objectives** of ProMembrane were:

- To identify and map the on-going research activities in the field of membrane technology in the Mediterranean region.
- To assess on-going research in membrane technology, identifying the gaps in knowledge and the barriers for the cooperation of R&D activities in a Mediterranean scope.
- To develop adequate recommendations for future research activities in order to fulfil the scientific and technological needs for the further implementation of membrane technologies in the Mediterranean region.

 ISBN: 9781843393702. Published by IWA Publishing, London, UK.

- To encourage young scientists in the membrane technology by the organisation of two regional scientific contests 'ProMembrane Young Scientist Award'.

The **principal outputs** of this SSA have been:

- The identification and mapping of the research organisations in the field of membrane technology, which focus on solving major problems with water supply and scarcity.
- A clear guideline of action regarding the future research and development in the field of membrane technology to overcome the needs of the Euro-Mediterranean countries.
- A project web-site and database platform to provide a conduit for international dissemination.
- Two regional ProMembrane Fairs in Syria and Egypt, which were the scenario for the regional researchers to present their work and results in posters and papers. Lectures were also given by members of the consortium.
- 'ProMembrane Young Scientist Award', a contest among researchers took place in each regional Fair, where the best six works were invited to the International Conference.
- The International ProMembrane Conference in Sfax-Tunisia, which gathered the most relevant actors in the membrane and water treatment field, such as research institutes and universities, manufacturers, SMEs, governmental authorities, end-users and specialists, to increase the awareness on research activities carried out in the Mediterranean Area, as well as the development of this technology to propose solutions to the water scarcity of the region.

As the main objective of ProMembrane was to encourage, support and focus the current research and development activities in membrane technology for water treatment in the Mediterranean area it seems relevant to further continue with this aim, to disseminate the project results. Therefore, the development of the already mentioned outputs will be exposed in the following chapters.

1.2 RELEVANCE TO THE OBJECTIVES OF THE INCO SPECIFIC PROGRAMME

INCO is the acronym used in the 6th Framework Programme for International Co-operation, which was implemented through three major routes:

- The opening of 'Focusing and Integrating Community Research' to third country organisations with substantial funding.
- Specific measures in support of international co-operation.
- International activities under the heading of Human Resources.

ProMembrane project is a clear example of supporting international co-operation, implemented thanks to the second tool mentioned above and more precisely through the specific programme 'Integrating and strengthening the European Research Area' (SP1).

ProMembrane addressed INCO objectives by:

- **Promoting international co-operation** in research and technology development in the field of membrane technology for water treatment and reuse.
- **Supporting the external relations** of the European Research community in reaching the EU water initiative and commitment towards Millennium Development Goals.
- Promoting close co-operation among the **European Research Area** and the Mediterranean Partner Countries.
- **Addressing environmental concerns** in the field of water resources and technological innovations for alternative water supplies, including membrane technology for the potable water supply and wastewater treatment and reuse.

1.2.1 International and Mediterranean Co-operation

As requested by the 2005 INCO work programme, ProMembrane aimed to integrate and strength the European Research Area, by supporting international co-operation in the field of water treatment and reuse in the Mediterranean Partner Countries by means of membrane technology. The principal objective of ProMembrane was to support the current research and development activities in membrane technology focussed on water treatment in the Mediterranean area, aiming the dissemination of research efforts and the exchange of knowledge and expertise, to point out the road for future research in the membrane field, in order to make the most of its technical and comparative advantages to further apply such technology to abate the water crisis in the Mediterranean basin.

Figure 1.1 shows the six flags of the Mediterranean countries involved in the project.

ProMembrane has created Euro-Mediterranean co-operation links among research institutes and universities devoted to the development of membrane technologies for the supply of potable water (desalination and treatment of surface water), as well as municipal and industrial wastewater treatment and reuse. It was also the scenario for European Institutions to strongly participate in international research initiatives in order to contribute to solve one of the major issues world-wide: water scarcity in the MENA countries. This has reinforced the human potential for

European research and has developed mutual beneficial research co-operation between Europe and third countries, in particular those from the Mediterranean Area.

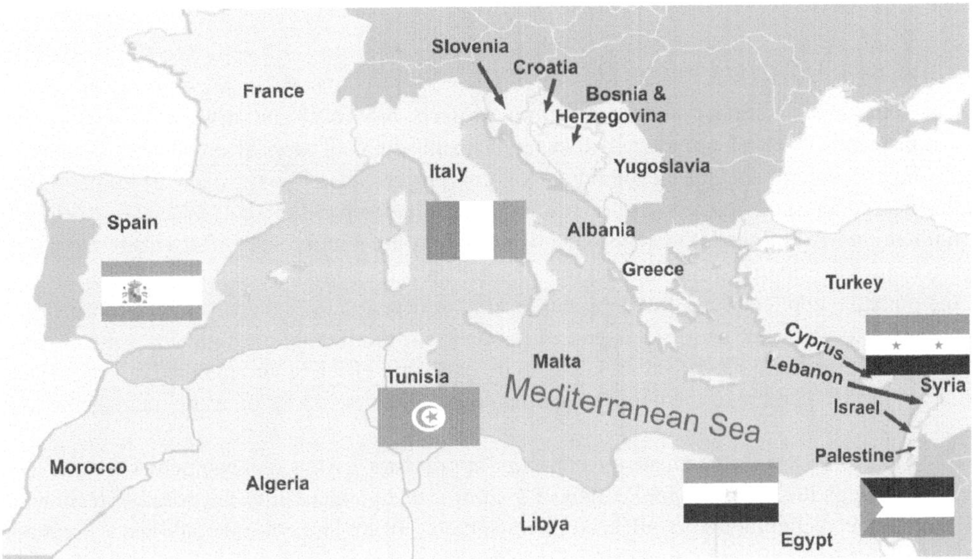

Figure 1.1 Mediterranean countries participants in the ProMembrane SSA

ProMembrane has reinforced European countries co-operation as it included professionals in the field of membrane technology for water treatment and reuse working in organisations and institutions in countries like **Germany, Italy and Spain**. It also has contributed to enhance co-operation among European countries and Middle East and North Africa countries like **Egypt, Tunisia, Syria and Palestine**. This project has given the opportunity to create an **Euro-Mediterranean Research and Innovation Area** in the field of membrane technology for water treatment and reuse, as a component of the opening of European Research Area towards this region. This has allowed the European research community to benefit from the know-how and expertise of MPCs in the field of membrane applications, enhancing the access to the knowledge elsewhere in the world.

ProMembrane promoted a better participation of third countries in addressing specific problems related to environment and sustainable development, as it is to abate the water crisis by the increased and efficient application of membrane technology for the supply of potable water and treatment of wastewater for reclamation and reuse. As a result of this co-operation with third country organisations, in particular those placed in the South and East of Mediterranean basin, a trans-national flow of knowledge, technological expertise and capacity building has been emphasised. This initiative has not only helped organisations from Mediterranean countries in their attempt to assist their countries in reaching a higher development status by means of having access to the latest technology and knowledge offered by membrane organisations from other countries with a higher investment on research, but also allowed the European research community to benefit from the expertise of developing country institutions acquired from the experience of dealing with a specific problem in a specific context.

1.2.2 INCO Mediterranean Partner Countries Specific Measures

ProMembrane has contributed to the **sustainable development** and socio-economic modernisation in the Mediterranean basin region, by developing the needed strategic alliances among research institutions to foster membrane applications in water treatment. The further application of membranes in different fields of water treatment would make water resources available to a greater number of people improving the way of life, following some of the main principles shared with some development aid policies like the EU Water Initiative and the Millennium Development Goals. On the other hand, this project has also helped to **implement European Water Framework Directive** as the promotion of membrane applications for the wastewater reuse contributes to the reduction of waste discharges to water streams decreasing their pollution, and achieving therefore a higher level of water quality. As one of the most important regional objectives, ProMembrane has **supported the economic integration of the two sides of the Mediterranean**, fostering socio-economic progress throughout this region and thus contributing to the objectives of the Barcelona Process.

ProMembrane filled in the scope of one of the main priorities of the INCO MED programme: the integrated management of limited water resources (B.1), which included a mix of such considerations as natural conditions, source of supply, technological considerations, socio-economic and socio-demographic conditions. Within this priority, the project refers to the three different areas of action:

B.1.1 Comprehensive Policy for Integrated Water Planning

Because most of the water treatment and reclamation projects are local initiatives, the experience acquired by Euro-Mediterranean members is seldom shared. Especially in the field of scientific research, as it is the case of the membrane technology development, the development of this project has shown a lack of cooperation, very limited know-how exchange and an uncoordinated use of resources, leading to parallel and ineffective R&D activities related to membrane applications. Additionally, the approach to water treatment and reuse differs from one country to another, and sometimes within one country, from one region to another. This lack of a holistic perspective regarding water treatment and the potential applicability of membranes has led at best to a very dispersed and confused system of water management, at worse to wasteful, or mismanaged investments in the membrane field for water treatment purposes. As the water resource is finite and its utilisation needs to be equitable efficient and planned, all regional strands should be interwoven, having the membrane techniques as a leading technology, key of water management for the twenty one century.

ProMembrane tried to find an agreement on common approaches to the management of water resources in the Mediterranean area through the effective application of membranes, by elucidating the possible technical solutions that can approach membrane technologies to all sectors of society, from large municipal water treatment plants and vast desalinisation plants to small scale modules for rural and peripheral areas, taking into consideration the different physical, socio-economical and cultural contexts that can be found in the Mediterranean basin. The unification of efforts in the research field in membrane technology for water treatment would provide the needed synergistic tools to further develop a technology which potentially overcomes the water crisis in the Mediterranean basin. Through the use of membranes it is possible to achieve the hygienic and chemical standards to reuse and reclaim effluents, closing the needed loop to ensure water sustainability in the Mediterranean region.

B.1.2 Improving the Water Consumption by Users and Uses and Plant Breeding for Efficient Water and Nutrient Use

Membrane technology represents the most efficient technical option to achieve the needed hygienic standards for the sustainable reuse of wastewater for irrigation in agriculture. Through its diverse techniques (micro, ultra and nanofiltration and reverse osmosis), membranes are capable to obtain an effluent that does not pose a risk to the health of the end users of the growing crops, by removing completely virus, bacteria and protozoa, as well as other dissolved components that threat human health. Among the different objectives covered by ProMembrane, it was also a priority to unify research efforts in the field of membrane for water reclamation bringing together experts that could contribute in the research of sustainable irrigation by the use of membranes.

B.1.3 Advanced Water Treatment, Re-use and Energy Implications

Membrane technology is considered an advance water treatment, in which different quality standards can be reached depending upon the regional and local needs. Through a wider membrane application, it is expected to achieve the needed mechanisms for water sustainability in areas suffering of water scarcity. ProMembrane has disseminated research efforts on membrane technologies for water and wastewater treatment, including municipal, industrial, brackish and surface water.

ProMembrane took into account the European Water Framework Directive, especially emphasising the applicability of membranes accomplishing public health standards. It also served to elucidate the research needs and requirements of this technology to be broadly applied in the Mediterranean region, taking into account economical, local, environmental, social and political factors.

1.3 THE CONSORTIUM

In order to run a highly efficient and successful project of this type, careful consideration was given to the composition of the project consortium. The proposed consortium was well balanced and allowed the maximum integration of knowledge, expertise and experience between partners. Members representing the **different areas of membrane technology for water treatment in the Mediterranean region** were part of ProMembrane, as it is shown in Figure 1.2.

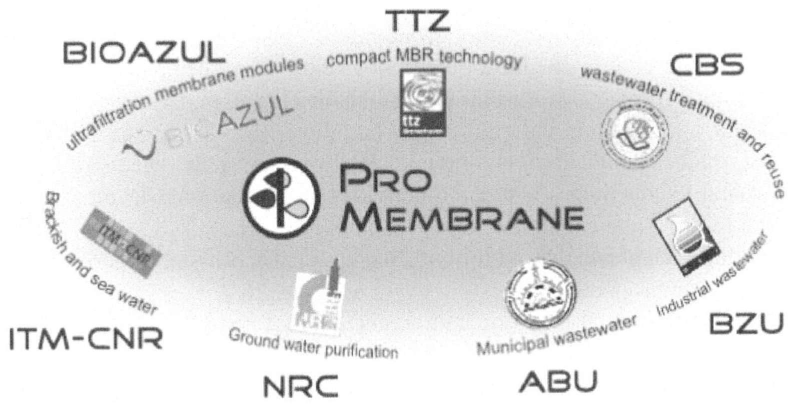

Figure 1.2 ProMembrane Consortium

1.3.1 Description of Consortium

1. Technologie Transfer Zentrum Bremerhaven (ttz), Germany

ttz Bremerhaven

ttz-Bremerhaven was established in 1987 and comprises six institutes for applied research and development (R&D). More than 100 highly qualified employees (scientists and engineers) are working on business-related projects in the several technological fields. All ttz departments have the most modern technical and laboratory infrastructure at their disposal. ttz has successfully carried out more than 4000 projects.

ttz works in many national and international projects in the field of membrane technology from which the following can be pointed out: MOWOM (mobile wastewater treatment process for olive mills, EU CRAFT project), IWAPIL (compact submerged membrane system for wastewater treatment applications for isolated locations, EU CRAFT), PURATREAT (energy efficient approach to the operation of membrane bioreactors for decentralized wastewater treatment, EU INCO-MED STREP), Bilge Water Treatment (treatment of bilge water onboard ships, German national project), ETHAFERM (continuous ethanol fermentation in a MBR system, German national project). Furthermore, membrane and wastewater treatment experts of ttz provide consulting services to national and international clients and institutions. Main goal of ttz's membrane activities in the field of water treatment is the improvement of water purification (regarding water quality and energy efficiency) to such an extent, that the purified wastewater can be re-used for technical purposes, for irrigation in agriculture or even for production of potable water.

2. Centre de Biotechnologie de Sfax (CBS), Tunisia

CBS is a governmental institution established in Sfax since 1989, specialized in the fields of environmental, industrial and agricultural biotechnology. CBS's mission is to develop the understanding and safe application of biotechnology through research and capacity building for sustainable development. CBS is one of the leading institutions in the field of environmental biotechnology in the African and MENA countries; developing and implementing new concepts for bioremediation of waste and wastewater, bio-energy production, biomass, enzyme production, conversion and use. Amongst four laboratories, the Laboratory of Environmental Bioprocesses has the role of studying the biological system involved in the aerobic and anaerobic degradation of urban and industrial pollutants. This laboratory had made experience in bioprocesses development such as fungal bioremediation of xenobiotics, bioremediation by extremophiles, biogas production, biological pre-treatment of industrial and urban effluents and membrane bioreactor technology. This laboratory has made experience in pilot test trials through his participation in several European projects.

3. Istituto per la Technologia delle Membrane-National Research Council (ITM-CNR), Italy

ITM at University of Calabria

ITM is the Institute of the Italian National Research Council (CNR) operating at national and international level within the framework of projects focused on Membrane Science and Technology. The Institute Headquarter is located at the University of Calabria, Rende (CS) with a unit at the University of Padova.

The multidisciplinary staff with diversified background is composed of 26 permanent workers, including the Director Dr. Lidietta Giorno, and about 40 temporary workers, i.e., visiting professors, associate professors, Ph.D. students, postdoctoral fellowships, and high-educational fellowships. The different backgrounds are in chemical engineering, process engineering, chemistry, biological science, food science, material science and physics. Significant exchange of young researchers, coming from various foreign Countries, e.g. Spain, The Netherlands, Russia, Algeria, South Korea, Japan, China, India, Bangladesh etc, has been always strongly encouraged in order to accelerate the growth of the structure and to integrate all research activities at international level. At present 10−15 external researchers per year visit the Institute and collaborate to its research programs.

Formal bilateral and multilateral agreements with South Korea, China, Russia, Algeria, Belgium, Turkey, Egypt, etc have been realized and/or are in progress.

The Institute is also involved in the sponsorization and organization of many different Conferences and Workshops (e.g. EMCC5 in 2008, EUROMEMBRANE in 2006, etc.)

In the last years, ITM-CNR has been involved in research projects, both national and international, co-funded by EU and private companies, on water desalination, water and wastewater treatment, mathematical modelling and simulation of catalytic membrane reactors, membrane contactors, and integrated membrane processes artificial hybrid organs and, more in general, on membrane operations.

4. BioAzul S. L.– R&D Engineering (BioAzul), Spain

BIOAZUL is an engineering company focusing on wastewater treatment and water reuse concepts. The product line of BIOAZUL includes complete SBR and MBR systems for wastewater treatment and water reuse especially in remote locations or for specific wastewater compositions. Other products (chemicals for wastewater treatment) and services are focusing on standard wastewater systems. The highly qualified staff has extensive experience in the treatment of specific wastewater and reuse processes (e.g. municipal, olive oil or paper industry).

Due to the participation in several EU R&D projects and own R&D activities in the fields of wastewater and solid wastes management, renewable energy and environmentally-sound agriculture, BIOAZUL has also gained in-depth knowledge of EU project management.

BIOAZUL's activities are divided in 3 interconnected business areas: R&D and commercialisation of own products, commercialisation of 3rd parties' highly innovative products, and consulting and management of R&D projects.

BIOAZUL has acted as partner and administrative coordinator of several European projects (9 environmental projects) within the 6FP and will be the administrative coordinator of one coordination action under the 7PF. In addition, technical activities have been carried out by the staff of BIOAZUL related to wastewater (treatments of municipal and industrial using different technologies), food wastes (solid and liquid mainly coming from agro-food industry), biomass production (for energy purposes) and soil remediation. Due to the active role of BIOAZUL participating in international (Europe, Latin America, Africa, Former Soviet Union) and national projects, BIOAZUL has an intensive network of contacts, including the olive market, and a good understanding of the situation with regards to current research activities in their fields of expertise. With regards to membranes, BIOAZUL participates/has participated in four EU projects:

- ALGATEC: general coordinator. The project is based on a photo-biorecator combined with a membrane module to treat washing water from the olive oil production. The project started in June 2009, will run for 24 months and two pilot plants will be installed and test in two cooperatives in Spain and Greece.
- IWAPIL: the project was based on a Membrane Bioreactor to treat wastewater coming from camping sites. One of the sites was located in Granada and BIOAZUL carried out technical activities in the staring and optimisation of the prototype. In addition, BIOAZUL got the rights on the MBR prototype and run it for one year after the finalisation of the project. The prototype is being tested currently in a hotel under the control of BIOAZUL.
- PROMEMBRANE: the main aim of the project was to promote and focus research activities on membrane technologies for the treatment of wastewaters in the Mediterranean Region.
- PURATREAT: within the project, 3 MBRs prototypes located in Tunisia have been tested using different membranes as promising alternative to conventional wastewater treatments. BIOAZUL was involved, among other tasks, in the evaluation of the performance of the MBRs.

5. National Research Centre (NRC), Egypt

The National Research Centre (NRC) was established as an independent public, governmental non-profit organization in 1956, with the aim 'to foster basic and applied scientific research, particularly in industry, agriculture, public health and other sectors of national economy'. NRC is the largest research centre for science and technology in the Middle East to serve the national, regional and international cooperation including technology transfer and membrane technology. NRC consists of

14 inter-disciplinary research institutes devoted to different R&D projects; organisation of conventions and symposia, as well as education, formation NRC conducted more than 5,450 successful R&D projects since its foundation. NRC has over 3250 highly skilled academic full-time employees (professors, Ph.D. holders, scientists and engineers). The annual turnover ranged from 12.3 to 18.7 million Euros. NRC includes several technical pilot plants for water, municipal and industrial scale processes, soil, agriculture, experimental large land areas as well as well equipped laboratories. One of the NRC large Institutes is the Environmental Research (ERI) with over than 150 Ph.D. holders, beside assistances and Post-Graduate researchers. The ERI has successfully undertaken several national, regional and international projects funded by Egypt, EU, USA, Canada and Germany.

6. Al-Baath University (ABU), Syria

Al-Baath University, one of the Syrian governmental universities, was established in 1973 and its Faculties are located in Homs and Hama, in the central region of Syria, where there are livestock, chemical and textile industries that in addition to the tourism, require the university attention to various types of science. Al-Baath University work to prepare graduates who are able to meet the needs of society, and to develop scientific research, especially those researches related to national issues relevant to economic and social development, humanitarian and technical. It also works to participate in the enrichment of human knowledge and global culture, nationalism, and address the strengthening of Arab identity appropriate to this age through the production of scientific, intellectual, literary and technical support for university students and teachers.

Al-Baath University is one of the four government's universities in Syria and has 20 faculties. One of the fourteen laboratories of the chemical engineering Faculty belongs to the environmental studies, where research in the field of water treatment is carried out, additionally a highly advanced pilot-plant for biological treatment of sewage water is installed.

7. Birzeit University (BZU), West Bank and Gaza Strip

Due to increasing water and environmental problems in the Middle East in general, and in Palestine in particular, Birzeit University has established the Institute of Environmental and Water Studies (IEWS). Since its creation in 2001, the IEWS is promoting multi-disciplinary R&D in sanitary and environmental engineering, waste and water resources management, and water science and technology. IEWS laid the foundation for the comprehensive M.Sc. programs (Water and Environmental Engineering, and Sciences) to meet increasing local and regional demands for qualified professionals in water and environmental engineering and sciences. The IEWS comprises 10 full-time faculty staff members, 5 contract and research staff from BZU and outside and 50 MSc students. Through engagement in more than 40 joint R&D projects, IEWS has partnerships with international collaborators and joined several EU programs on capacity building, environmental education and wastewater treatment and reuse. IEWS has a full array of advanced analytical instrumentation within its laboratories, and access to additional specialized facilities in other departments within Birzeit University. BZU has a central WWTP with zero-liquid discharge, training and research facility with several onsite treatment systems.

1.4 WORKPLAN

The project ran for two years, from the 15th of August 2006 to the 14th of August 2008. The content and structure of the work plan was designed to achieve the project aims as efficiently as possible. The first stage of the project covered the identification, mapping and evaluation of the on-going research, with the objective of proposing future research strategies which helped to overcome the current technical barriers of application. The second stage was the diffusion and dissemination of the successful experiences and research activities being carried out in that moment, through the organisation of regional fairs in targeted regions and an international conference, in order to encourage the further research activities in membrane technologies. The diverse activities and tasks were divided into five Work Packages (WPs), associated with a key activity and coordinated by a work package leader. They were divided as follow:

- **Work Package 1** – Project management (WP1) was devoted to the management of the project, and was responsible for all of the other work packages. In order to secure an efficient running of the project, the management tasks were divided into three types of activities: General co-ordination and Financial Project Management, Administrative Project Management and Scientific Project Management. WP1 included the material and human resources management, as well as the establishment of the project's website, communication to the partners of the project's progress and their obligations and the delivery of obligatory reports to the Commission.
- **Work Package 2** – Mapping Current Research Exercise (WP2) started at the beginning of the project and acted as basis for the rest of the work. It aimed to collect relevant information about current research being

performed in the field of membrane technology as a water treatment method in the Mediterranean basin countries. The work was divided in two geographical regions, Middle East (region A) and North Africa (Region B), where the members of the consortium identified the centres of excellence and research centres. Once they were geographically mapped, a formation of research clusters regarding field was proposed, including (1) municipal and (2) industrial wastewater treatment, (3) surface water purification and (4) brackish and sea water treatment for drinking purpose. The information was compiled and transferred to Work Package 3 and 5 where it was employed for further activities.

- **Work Package 3** – Assessment of on-going research (WP3) processed the information gathered in previous step to evaluate the state of the research on membrane technology in order to identify needs, gaps in knowledge, possible cooperation, overlapping and the potential application in MPCs regional conditions. The development of this work package was a key figure in completing ProMembrane goals, since it was the scene which allowed discussion and evaluation of the scientific and technological needs of this technology to be further applied as a sustainable method for the water crisis of the Mediterranean basis. A list of needs and requirements for the membrane technology in order to be further applied in the Mediterranean area was compiled and analyzed in the following chapters.
- **Work Package 4** – Development of research and development strategies (WP4) contained ProMembrane the project's research work most important deliverable: the research and development strategy. In this work package, recommendations for the future research and development were made, regarding cooperation strategies, by identifying and shaping Euro-Mediterranean research clusters in membrane technology for water treatment, and scientific strategies, by pointing out the hot points and the weakness that had to be overcomed.
- **Work Package 5** – Promotion and diffusion of membrane technology research in the Mediterranean Area (WP5) was completely devoted to the diffusion and dissemination of the successful experiences and research activities being carried out during the project development, in order to encourage the further research activities in membrane technologies. The dissemination plan included four main tools: (1) ProMembrane Regional Fairs: Region A in Homs-Syria (congregating the Middle East countries) and Region B in Cairo-Egypt (congregating the North African region); (2) International ProMembrane Conference in Sfax-Tunisia; (3) ProMembrane website and the (4) ProMembrane Conference Proceedings Publication. This work package was responsible to establish joint strategies and foundations to ensure future collaboration and the continuity and self-funding of the network after the project time.

The relationship between the work packages is demonstrated graphically in the Pert diagram below:

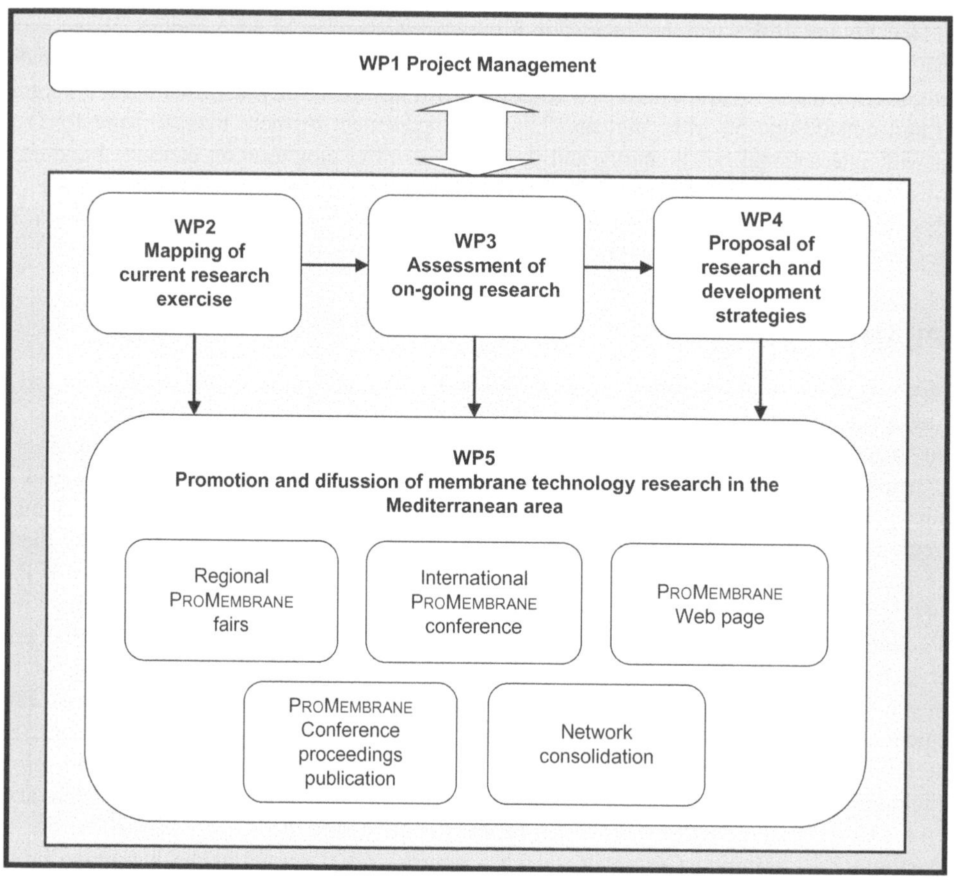

Figure 1.3 ProMembrane Pert diagram

The following Gantt chart shows the distribution of the work within the project into the different work packages and tasks, which shows the number of the months and WPs and tasks duration. ProMembrane consortium meetings are marked by a triangle. The most important deliverables are marked by a flag and these mainly represent the online updates of information, project reports and important deliverables such as the flyers and the program definition for the ProMembrane Regional Fairs and Conference. The milestones are represented by starts, which indicate the most important project achievements.

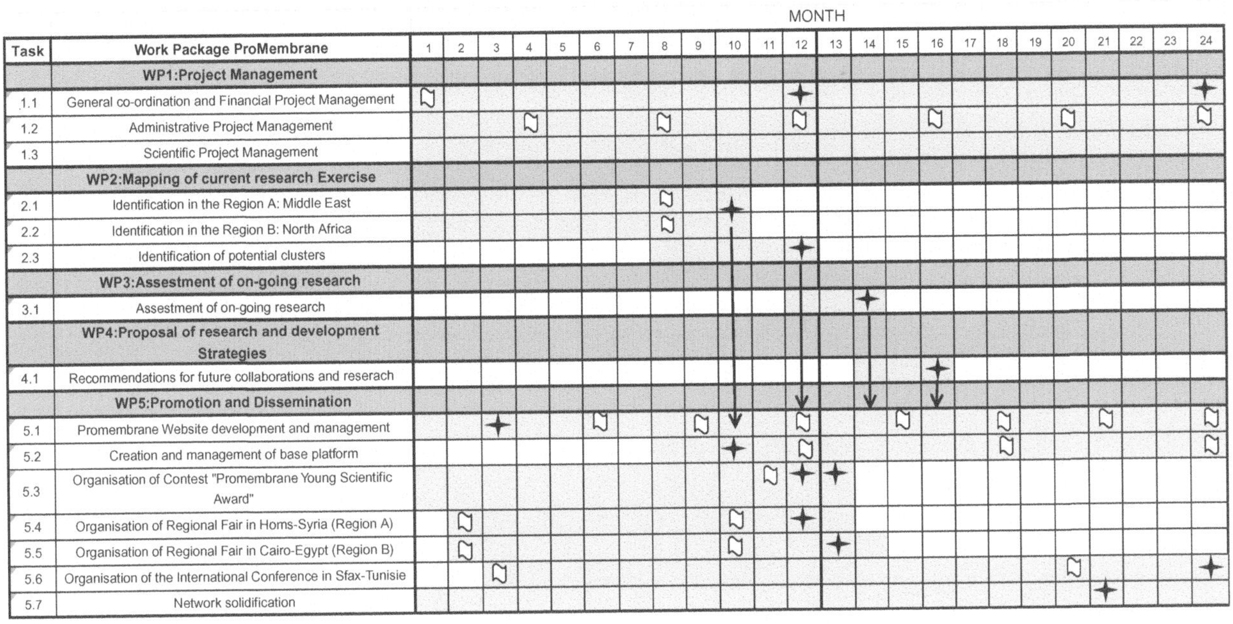

Figure 1.4 ProMembrane Gantt diagram

Chapter 2

Identification and knowledge mapping of the research organizations in the field of MT

Rashed Al-Sa`ed, Hussein Abdel-Shafy, Fouad Atallah, Ziad Mimi, Sami Sayadi, Alessandra Criscuoli and Gerhard Schories

INTRODUCTION

In this chapter, we first discuss various aspects of current research efforts within the MENA countries focusing on current available published data. Secondly, using membrane technology (MT) in wastewater management, certain gaps of knowledge and overlapping activities are identified. Finally, recommendations pertinent to future research on MT are suggested to enhance their implementation as advanced sustainable water treatment processes.

The findings of this chapter are based on authors' own experience; data collected from scholar literature (114 articles of 550 peer-reviewed papers) on the application of MT for both water and wastewater treatment, and 22 installed lab and full scale membrane based treatment systems. To compile data on MBR installations and published works from the countries under study, the WERF-Database and MBR-Network were accessed (http://www.mbr-network.eu/mbr-database/literature.php).

National R&D programs for the MBR technology applications in wastewater treatment and reclamation have started in some Arab MENA countries, such as Algeria, Egypt, Morocco and Tunisia. Some of the research efforts made are presented and discussed.

2.1 EXPERTS AND TOPICS WORKING IN THE FIELD OF MEMBRANE TECHNOLOGY

In Palestine, a total number of 80 researchers identified their field of expertise in membrane technology including desalination application in water and sanitation facilities. Figure 2.1 illustrates the distribution of the researchers on R&D institutions, where about 67% are working in both academic and governmental agencies. The NGO (non-governmental organization) sector is actively working in the field of MT, where about 22% of experts are engaged in both local and foreign NGOs. The industrial sector has about 12% of experts employed. Almost half of the Jordanian academic staff (55%) are professionals working in the governmental sector (25%), leading to 80% of the experts are working by the public sector. Compared to Jordan and Palestine, the situation in Lebanon is totally different, where the vast percentage (60%) of membrane professionals is employed by the industry. Despite many attempts to collect technical data on academic and governmental experts, only few were identified. About 24% of MT experts were found working at academic institutions and even less (4%) are engaged in governmental departments.

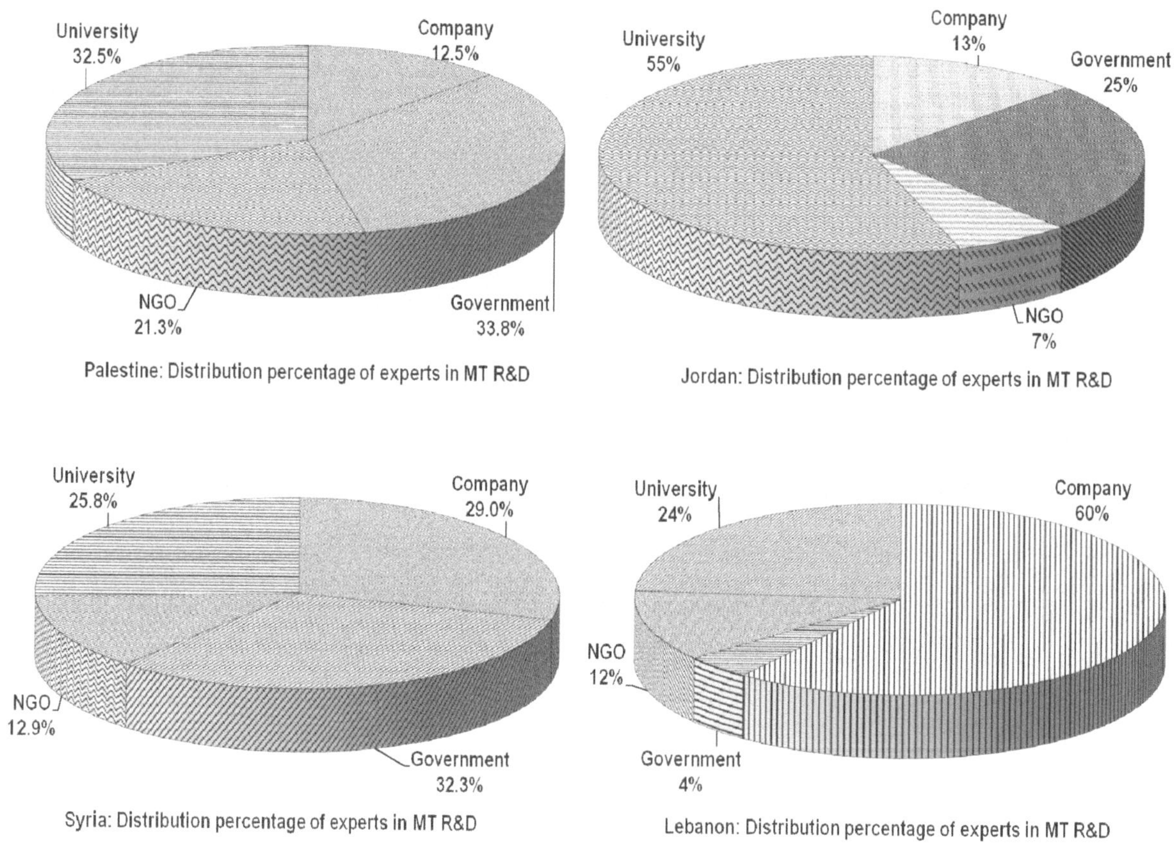

Figure 2.1 Distribution of professionals identified as MT experts in region A

For Syria, Figure 2.1 shows a similar percentage of expert's distribution working at academic departments, however, with fewer shares (29%) of industry involved in MT technology planning and marketing. It is worth mentioning that a total percentage of 58% of MT identified themselves are employed within the public sector (academia and government), compared with 80 and 67% in Jordan and Palestine, respectively.

Compared with compiled published work from North Africa (Region B: total 70 articles), the total publications of 44 mapped in Area A were grouped into the following application areas (1) water treatment for potable purposes; (2) wastewater treatment (domestic, grey wastewater, industrial, municipal, leachate and effluent reclamation). The grouping was performed according to the main objective of the PROMEMBRANE project. As the number of professionals specialized in MT is limited in Region A, and the region is characterized as water scarce zone, the main focus of publications was on use of membrane technology applications for water treatment (Figure 2.2). During the five years (1996–2000) very few papers (6) tackled the MT for water and wastewater treatment. However, Figure 2.2, where the number of published papers on water treatment (both brackish and saline water) is more than those for application of MT for wastewater treatment, 29 and 15 respectively.

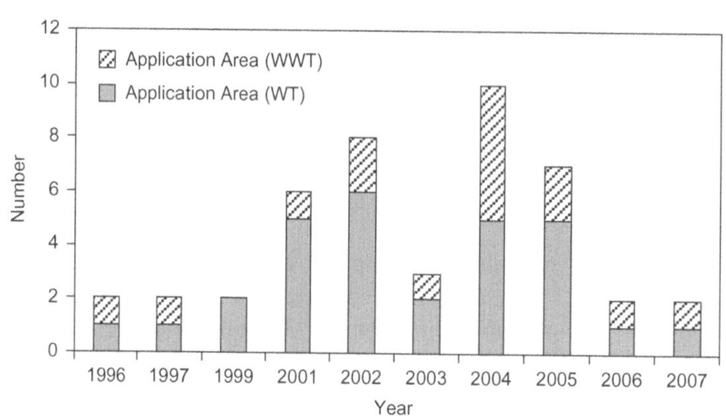

Figure 2.2 Research publications on application areas of MT technology in region A

A rapid increase of published articles is identified during the five years (2001–2005), where about 52% of papers were on use of MT for both water desalination of brackish water and marine water sources for drinking water purposes. Also a marked increase 25% of published scientific work is made on use of MT in wastewater treatment and effluent reclamation. This shows, definitively, the importance of membrane processes as a part of water scarcity and production of reclaimed effluent suitable for agricultural irrigation. This might be induced by annual drought periods, limited quantity and degraded quality of available freshwater resources due to overexploitation, additions of allocthonus pollution loads and salt intrusions, stringent local and regional standards on treated effluent destined to agricultural irrigation. Also, the current professional batch within this region come directly from either M.Sc. or Ph.D. programs with little practical experience in engineering offices or MT companies working at international level.

Compared to membrane technology applications in water treatment (desalination), the introduction of wastewater membrane and reclamation facilities in the MENA region has been if any, more regionally based. While in the Mediterranean zone (region A), a larger number of research studies were conducted in the area of potable water desalination than for wastewater treatment, the situation was reversed in North Africa (region B). This could be attributed to the existence of well established research groups, funds availability, and enforcement of national laws pertinent to protection of tourist coastal areas. This enabled MBR research and applications to flourish at lab, pilot and full scale installations for high strength industrial flows, municipal sewage treatment and tourist sites.

It is assumed that the caution of North African municipalities to consider advanced treatment systems as MBRs to the well established conventional treatment options, might have delayed the application of MBRs into the municipal arena. However, the promotion of MBR in industrial applications, particularly for high strength, difficult to treat waste streams, allowed for alternative technologies such as MBRs. This has been reflected in more than 70 published articles from research groups in Tunisia, Algeria, and Morocco, and Egypt the installation of three full scale installations for industrial wastewater treatment in Tunis, and Algeria (Ben Aim and Semmens, 2003; Bentama *et al.*, 2004; Benzaoui and Bouabdallh, 2004, Abdel Shafy *et al.*, 2005; Hussaine-Sadi and Sadoun, 2005). As per May 2007, the total number of installed membrane based wastewater treatment plants (immersed and external MBRs, UF/RO) in lab and full scale applications (domestic, industrial and municipal) has reached 22 (Table 2.1).

Table 2.1 Number of membrane based WWTPs in the Mediterranean and North Africa areas

Number	Full scale	Country	Number	Lab scale	Country
3	Domestic	Egypt	0	Domestic	Non
3	Municipal	Palestine, Israel	4	Municipal	Tunis
3	Industrial	Tunis, Algeria	9	Industrial	Tunis, Morocco, Algeria
Total number	9		13		

More than half of the membrane based WWTPs installed within the MENA zone is installed for R&D activities including membrane development, process performance optimization, and fundamental research. Table 1 shows that all installed membrane based treatment systems, except the one (UF/RO) applied as a post-treatment stage, are of MBR type (Quteishat *et al.*, 2003; Khamis *et al.*, 2005; Oron *et al.*, 2008; Al-Sa`ed *et al.*, 2009). The MF modules are applied for both municipal and industrial wastewater treatment. Eight out of 12 of the full scale installed MBRs, of which 50% serve municipal sector, are membrane immersed with hollow-fiber configuration, polymeric material and ZeeWeed fabric products.

Compared to conventional WWT systems, MBRs are advantageous as of small land area demand, process flexibility and excellent reclaimed effluent which suitable for various purposes including unrestricted agricultural irrigation. However, MBRs show higher energy demand, require a higher level of automation, skilled operational staff and frequent cleaning due to fouling and scaling (Al-Mutaz, 1991; Wang *et al.*, 2005; Younos, 2005; Le-Clech *et al.*, 2006; Yang *et al.*, 2006; González *et al.*, 2007). To-date, much of the research efforts made within the African countries (region B) on membrane based treatment systems have mainly focused on bench or pilot scale studies (Abdessemed and Nezzal, 2002; Abdel-Shafy *et al.*, 2005; Abdessemed and Nezzal, 2005; Khemekhem *et al.*, 2006) and short term operations in municipal applications. Lab scale studies on industrial applications with particularly high strength and difficult to treat waste streams, however, were conducted as alternative technologies such as MF/UF/NF (Bouguecha and Dhahbi, 2002; Saffaj *et al.*, 2004; Fersi *et al.*, 2005; Masmoudi *et al.*, 2006; Bentama *et al.*, 2007). However, regardless of the source of wastewater, whether it is municipal or industrial, very few publications involved full-scale studies for long-term operational periods (Khider *et al.*, 2004; Dhouib *et al.*, 2006). For pathogen removal and municipal wastewater pre-treatment, anaerobic MBRs have proven to be a particularly reliable technique (Ellouze *et al.*, 2006; Saddoud *et al.*, 2007). In Region A, few research studies (Harussi *et al.*, 2001; El Sheikh *et al.*, 2003; Quteishat *et al.*, 2003; Khamis *et al.*, 2005; Balkwill and Arviv, 2006; Al-Sa`ed *et al.*, 2008; Oron *et al.*, 2008) were made on development and use of local material to improve membrane structure, control fouling, and improve flux permeate at lab scale experiments using MF/UF membranes.

Figure 2.3 illustrates one of the databases developed using Excel program, where the data (only scholar literature between 1996–2007) can be filtered (indexed) according to users' search. The table below shows an excerpt of articles search pertaining to the use of MT only for wastewater treatment (WWT) in Arab MENA region (country filtered).

Authors/Names	Year	Title	Journal Name, volume (issue), pp	Abstract	Country	Technology used	USE (WT/WWT)
M.Khamis,A.Manassra,M.Dakiky	2005	Wastewater treatment by advanced membrane Technol	International conference:water values	in order to obtain effluent that can be used	Palestine	UF-RO	WWT
Badania,Z., Ait-Amara,H., Si-Salahb,A	2005	Treatment of textile waste water by membrane bioreacto	Desalination 185 (1-3) 411–417	The objective of this study is to determine	Algeria	MBR	WWTdomestic
Mandil,M.A.,Bushnak,A.A.	2002	Future needs for desalination in South Mediterranean co	Desalination 152 (1-3) 15-l 8	This is an attempt to forecast the role des	Syria	Desalination	WT &WWT
Afonso,M.D., Maria,A., Alves,B., Moh	2002	Crossflow microfiltration of marble processing wastewat	Desalination 149 (1-3) 153–162	MF is an efficient and ecologically suited e	Jordan	MF	WWT/Domestic
Saddoud,A., Hassaїri,I., Sayadi,S.	2006	Anaerobic membrane reactor with phase separation for	Bioresource Technology xxx (2006) x	This study examined the feasibility of appl	Tunisia	Cross-flow mic	Cheese whey
Dhouib,A., Hamad,N., Hassaїri,L.,	2003	Degradation of anionic surfactants by Citrobacter braaki	Process Biochemistry 38 (2003) 1245	To select and isolate bacteria or consortia	Tunisia	Membrane bio	WWT
Sayadi,S., Allouche,N., Jaoua,M., Al	2000	Detrimental effects of high molecular-mass polyphenols	Process Biochemistry 35 (2000) 725–	To investigate whether low, medium or hig	Tunisia	UF	WWT/(OMW)
Dhouib,A., Hdiji,N., Hassaїri,I., Saya	2005	Large scale application of membrane bioreactor technol	Process Biochemistry 40 (2005) 2715	A powerful Citrobacter braakii strain able	Tunisia	(UF)Membran	WWT/Industrial
Abu Zahra,B.A.A.	2001	Water crisis in Palestine	Desalination 136 (1-3) 93–99	Palestine is experiencing a severe water c	Palestine	Desalination	W&WWT
Ahmed,S.A., Tewfik,S.R.,Talaat,H.A.	2002	Development and verification of a decision support syste	Desalination 152 (1-3) 339–352	This paper addresses the development an	Egypt	RO	WWT/domestic
Masmoudi,S., Larbot,A., El Feki,H. B	2006	Elaboration and properties of new ceramic microfiltration	Desalination 190 (1-3) 89–103	In this work, we were interested in the dev	Tunisia	ceramic micro	WWT/Industrial
Khemakhem,S., Larbot,A., Ben Amar	2006	Study of performances of ceramic microfiltration membr	Desalination 200 (1-3) 307–309	In this study a microfiltration membrane fr	Tunisia	MF	WWT
Ammary,B.Y.	2005	Treatment of olive mill wastewater using an anaerobic s	Desalination 177 (1-3) 157-165	Wastewater from olive mills in Jordan has	Jordan	lab scale ana	olive mill WW
Fersi,C., Gzara,L., Dhahbi,M.	2005	Treatment of textile effluents by membrane technologies	Desalination 185 (1-3) 399–409	This paper presents the results of the labo	Tunisia	MF,UF,NF	WWT/Industrial
Abdel-Shafy,H.I., Hegemann,W., Gui	2005	Membrane Bioreactor for the Treatment of Wastewater	CEJOEM 2005, Vol.11. No.3.: 217–22	The membrane bioreactors (MBRs) are st	Egypt	MBR	WWT/Municipal
Al-Qodah,Z., Shawaqfeh,A.T., Lafia,V	2007	Adsorption of pesticides from aqueous solutions using o	Desalination 208 (1-3) 294–305	Adsorption characteristics of the pesticide	Jordan	oil shale ash (	pesticides
Bindra,S.P., Muntasser,M., El Khweld	2003	Water use efficiency for industrial development in Libya	Desalination, Volume 158, Issues 1-3	desalination are presented. The paper is c	Libya	desalination	WWT/Industrial
Bremere,I., Kennedy,M., Stikker, A.,	2001	How water scarcity will effect the growth in the desalinat	Desalination, Volume 138, Issues 1-3	Desalination technology is finding new out	Libya	Desalination	WWT/Municipal

Figure 2.3 Database of published scholar literature on MT application for wastewater treatment in Arab MENA region

2.2 KNOWLEDGE GAPS AND OVERLAPPING ACTIVITIES IN MT

In a similar way, the 44 total publications found in region A, were classified and grouped into the following main research fields (1) general research (GR) papers on review and theoretical aspects of MT use in water treatment and wastewater treatment; (2) fundamental research aspects; which includes operation and design parameters, fouling, cost, modelling, membrane aided treatment systems, hybrid modules, pre- and post-treatment. As the number of research papers is few, the papers were grouped in only two application areas without differentiation on the type of water or wastewater. Figure 2.4 illustrates chronically the distribution of published papers within the two research topics of MT applications; general research issues (GR) and specialized applied research themes (FR).

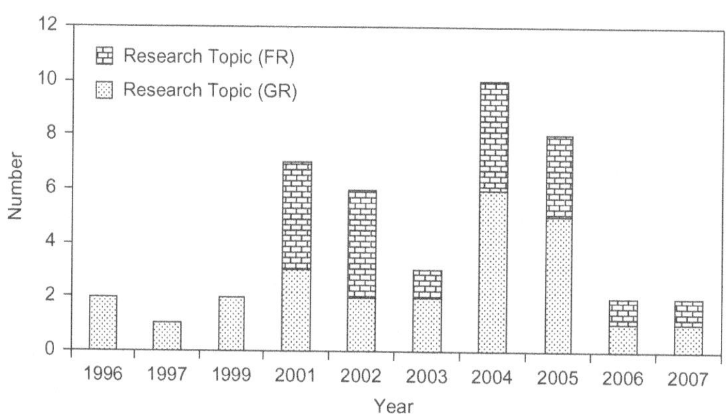

Figure 2.4 Research publications on application areas of MT technology in region A

It is obvious that only general research papers (GR) on MT applications are made during the nineties (1996–1999), this is clearly due to the limited human and financial resources available in the region. However, the applied research (FR) has been published from conducted research on MT application in both water and wastewater treatment first in 2000. There are sharp variations in the number of published work among the individual countries with the Area A. For example, only 4 papers were found published by the experts in both Lebanon and Syria, while Palestine and Jordan have published 10 and 29 respectively. GR papers constitute about 57% while applied fundamental research (FR) entailed only 41% of the total published articles in this region. This is a good sign that MT has gained interest from the research community and policy makers for the provision of drinking water and improving effluent quality.

2.3 TECHNICAL REQUIREMENTS AND NEEDS FOR COOPERATION

The results obtained identified wide research areas related to the types of scientific and technical advances that are crucial for membrane water treatment technologies to find broad acceptance and application in the countries under study. To achieve sustainable, affordable and adequate wastewater treatment facilities in the MENA zone, allocation of regional research funds that will guide the membrane processes research in the Mediterranean and North Africa region are required.

It was demonstrated that the availability of experts and well-trained practitioners, well-equipped research facilities, availability of funds, economical affordability and technical feasibility of MT, as well as the official commitment and endorsement of stringent effluent standards on treated effluent for intended uses depend on each other.

Protecting the quality of surface and groundwater requires sustainable management on a watershed basin scale to consider all impacts on the water. Of equal importance, public health protection calls for safe brine disposal or effluent reuse (Nghiem *et al.*, 2004; Wilderer, 2005; Galil and Levinsky, 2007). This implies use of advanced treatment technologies, among of which are membrane based processes. However, because water quality criteria are dependent on local conditions within the MENA region, it is necessary to define groups of similar rivers by clustering them as eco-regions. As protection of the aquatic environment and public health pertinent to emergent pollutants more often lacks a confirmed scientific, financial and managerial basis, further research and updating of criteria based on large scale MBRs are needed (Gander *et al.*, 2000; Chen *et al.*, 2003; Abdel-Shafy *et al.*, 2005; Cornel and Krause, 2006).

Forming cooperation between technology developers and leading companies is a crucial approach that dramatically shortens the time required to promote membrane technology and address some of key business problems and challenges. The approach provides ready access to professionals best qualified to provide commercial development guidance for new MT based options (Oron *et al.*, 2004; Oron *et al.*, 2008). It enables development or co-development of new membrane based systems through collaboration or joint development with trend setting companies who are expert and leaders in their fields (Escobar, 2005; Frechen *et al.*, 2006; Quazzani and Bentama, 2008).

Water scarcity in the MENA region has promoted the use of unconventional water sources, namely seawater, brackish water and treated/reclaimed wastewater, which are actually unlimited. With adequate membrane design, utilizing MF/UF as pre-treatment processes might reveal wastewater desalination costs lower than seawater desalination cost, thus making wastewater desalination as one of the feasible processes to produce water quality that fits many industrial water quality requirements (Adham *et al.*, 2001; Rebhun, 2004; Aboabboud and Elmasallati, 2007; Liao *et al.*, 2007; Wichelns and Nakao, 2007; Elhassadi, 2008). In the future it is likely that direct processing of wastewater in MBRs followed by RO will open up more opportunities for effluent reuse for a wide range of purposes. However, several studies (Bouhabila *et al.*, 2001; Fane and Fane 2005; Fane, 2007; Rosenberger *et al.*, 2006; Zhang *et al.*, 2006) made on MBRs sustainability, revealed an overall sustainability for the MBRs as good, however the current capital and operational costs of membrane based treatment technologies may not necessarily satisfy some economical and ecological sustainability criteria (Oron *et al.*, 2004; Ondrey, 2005; Le-Clech *et al.*, 2006; Yang *et al.*, 2006).

The technical feasibility of MT applications in wastewater treatment is very well documented, but the widespread utilization of membrane based processes is constrained by the high capital and operational costs. The price of membranes, their replacement frequency and the electrical energy consumed are the most important factors influencing the global costs of the processes (Semiat, 2000; El Sheikh *et al.*, 2003; Al-Bastaki, 2004; Kim *et al.*, 2004; Fane and Fane, 2005; Lesjean *et al.*, 2005; Saddoud *et al.*, 2007; Ellouze *et al.*, 2009). Thus, it is important to select an adequate membrane type as well as to optimize the operational conditions of the preceding treatment stage case by case. However, recent rapid proliferation of MBRs as a result of the technological advances and reduced costs has resulted in many owners, operators and engineers considering them as part of plant upgrades and expansion plans of overloaded sewage works (Oron *et al.*, 2004; DWA, 2005; Ondrey *et al.*, 2005; Balkwill and Arviv, 2006; Cornel *et al.*, 2006; Frechen *et al.*, 2006; Le-Clech *et al.*, 2006; Yang *et al.*, 2006; Quazzani and Bentama, 2008). Nevertheless, a breakthrough in MT advancement for wastewater treatment and effluent reclamation can be only made, if we can prove its cost effectiveness and sustainability for developing counties, a revolution in nanoscience and membrane engineering will have potential impacts on social, environmental and economical development as well as on the political stability in Arab MENA countries in particular.

2.4 CONCLUSIONS AND RECOMENDATIONS

Establishing databases on published literature and experts based on literature survey and distribution of questionnaires to collect reliable data on membrane technology applications from individual MENA countries was a challenge and faced with numerous obstacles. These include but are not limited to lack of funding, management commitment, as well as professional experts and practitioner personnel. Other barriers most commonly cited by respondents include:

- Absence of competitive industry and less economical motivations
- Lack of regulatory motivation at local and regional levels

- Un-coordinated training and lack of R&D programs
- Limited practical experience by academic and unskilled operating staff
- Lack of funds for both investment and operation

These barriers make the implementation of membrane technology for wastewater treatment and reclamation difficult, and impede their sustainability and wide-scale promotion in the MENA region. The current membrane based wastewater treatment systems are almost entirely MBR based technology at pilot-scale levels for industrial wastewater, and are predominantly located in the North African countries. Growth in membrane technologies in the municipal sector is likely to be advanced by a combination of decreasing process costs, increasing stringency of environmental legislation and further process innovation, such as the submerged membrane module. A breakthrough in the industrial commercialization of membrane based wastewater treatment technologies in the MENA region will only be possible by converting the MT into a profit-making position. This can be enhanced by close cooperation between industries, research communities and decision makers, utilizing of committed and reliable multidisciplinary research groups in various R&D fields. The following recommendations can be made:

- Building on the current efforts and recent technological advancements, a strategic research program should be developed to enhance the R&D activities pertinent to performance improvement and costs reduction of current membrane based wastewater treatment technologies in the MENA region.
- A strategic research investments program including budget estimates require an adequate and shared funding support from the industry, governmental, public and private sectors. The needed funds to implement this program will promote innovative research; enhance capacity building, award research efforts, advance knowledge transfer and communication.
- Establishing sustainable networking, initiation public-private-partnerships, creation of national, regional and international industrial alliances, conduction of public awareness campaigns, and creation local and regional R&D incubators are among the crucial efforts to enhance the promotion of membrane based wastewater treatment technologies in the MENA countries and worldwide.

2.5 REFERENCES

Abdel-Shafy H. I., Hegemann W., Guindi Kh.A., Tawfik N. S. and Teschner K. (2005). Membrane bioreactor for the treatment of wastewater in an Egyptian plant. *Central European J. Occup. Environ. Med.*, **11**(33), 217–223.

Abdessemed D. and Nezzal G. (2002). Treatment of primary effluent by coagulation-adsorption-ultrafiltration for reuse. *Desalination*, **152**, 367–373.

Abdessemed D. and Nezzal G. (2005). Tertiary treatment of a secondary effluent by the coupling of coagulation-adsorption-ultrafiltration for reuse. *Desalination*, **175**, 135–141.

Aboabboud A. and Elmasallati S. (2007). Potable water production from seawater by the reverse osmosis technique in Libya. *Desalination*, **203**, 119–133.

Adham S., Gagliardo P., Boulos L., Oppenheimer J. and Trussell R. (2001). Feasibility of the membrane bioreactor process for water reclamation. *Water Sci. Technol.*, **43**, 203–209.

Al-Bastaki N. M. (2004). Performance of advanced methods for treatment of wastewater: UV/TiO2, RO and UF. *Chem. Eng. Processing* **43**, 935–940.

Al-Mutaz I. S. (1991). Environmental impact of seawater desalination plants. *Environ. Monit. Assess.*, **16**, 75–84.

Al-Otaibi A. H. and Abdel-Jawad M. (2007). Water security for Kuwait. *Desalination*, **214**, 299–305.

Al-Sa`ed R., Khamis M. and Dakiky M. (2008). Assessment of current conventional and membrane technologies for wastewater treatment and effluent reclamation in Palestine. *Proceedings of the International Symposium on Biotechnology*, May 4–8, Sfax, Tunis, pp. 135–141.

Al-Sa`ed R., Sayadi S., Ghata A., Abdel-Shafy H., Schories G., Oropeza M., Lorenzo A. and Drioli E. (2009). Advancing membrane technologies for wastewater treatment and reclamation in selected Arab MENA countries. *Desalin. Treat.*, **4**, 287–293.

Badani Z., Ait-Amar H., Si-Salah A., Brik M. and Fuchs W. (2005). Treatment of textile waste water by membrane bioreactor and reuse. *Desalination*, **185**, 411–417.

Balkwill A. and Arviv R. (2006). Recent progress and development in UF tertiary treatment and MBR technology. *Proceedings of the International Conference on Innovations and Applications of Sea-Water and Marginal Water Desalination*. Israel Desalination Society, Haifa, Israel. http://gwri-ic.technion.ac.il (accessed 8 June 2009).

Ben Aim R. M. and Semmens M. J. (2003). Membrane bioreactors for wastewater treatment and reuse: a success story. *Water Sci. Technol.*, **47**, 1–5.

Bentama, J. Ouazzani K., Lakhliai Z. and Ayadi M. (2004). Inorganic membranes made of sintered clay for the treatment of biologically modified water. *Desalination*, **168**, 295–299.

Bentama J., El Ghzizal A. and Ferrandis J. Y. (2007). Ultrasonic waves for characterisation of clay membranes. *Desalination*, **206**, 1–8.

Benzaoui A. and Bouabdallah A. (2004). Desalination and biological wastewater treatment process. *Desalination*, **165**, 105–110.

Bouguecha S. and Dhahbi M. (2003). The role of membrane technologies in supplying drinking and industrial water in Tunisia: Conventional process and new trends. *Desalination*, **151**, 75–86.

Bouhabila E. H., Ben-Aim R. and Buissonz H. (2001). Fouling characterisation in membrane bioreactors. *Sep. Purif. Technol.*, **22/23**, 123–132.

Bushnak A. A. (1999). Water supply challenge in the Gulf region. *Desalination*, **78**, 133–145.
Cartwright. P. (2005). Membranes for process water reuse. *Chemical Engineering*, **111**, 38–42.
Chen T. K., Ni C. H., Chen J. N., Lin J. T. and Chang C.Y. (2003). Application of a membrane bioreactor system for opto-electronic industrial wastewater treatment – a pilot study. *Water Sci. Technol.*, **48**, 195–202.
Cornel P. and Krause S. (2006). Membrane bioreactors in industrial wastewater treatment–European experiences, examples and trends. *Water Sci. Technol.*, **53**, 37–44.
Dhouib A. Aloui F. Hamad N. and Sayadi S. (2006). Pilot-plant treatment of olive oil mill wastewaters by *Phanerochaete chrysosporium* coupled to anaerobic digestion and ultrafiltration. *Process Biochem.*, **41**, 159–167.
DWA, *German Association for Water* (2005). Wastewater and Waste e.V., Working Report No. 2, Technical Committee on MBR Technology at Municipal WWTPs, Hennef, Germany.
Elhassadi A. (2008). Horizons and future of water desalination in Libya. *Desalination*, **220**, 115–122.
Ellouze M., Saddoud A., Dhouib A. and Sayadi S. (2009). Assessment of the impact of excessive chemical additions to municipal wastewaters and comparison of three technologies in the removal performance of pathogens and toxicity. *Microbiol. Res.*, **164**, 138–148.
El Sheikh R., Ahmed M. and Hamdan S. (2003). Strategy of water desalination in the Gaza Strip. *Desalination*, **156**, 39–42.
Escobar I. C.(2005). Membrane developed systems for water and wastewater treatment. *Environ. Progress*, **24**, 355–357.
Fane A. G. (2007). Sustainability and membrane processing of wastewater for reuse. *Desalination*, **202**, 53–58.
Fane A. G. and Fane S. (2005). The role of membrane technology in sustainable decentralized wastewater systems. *Wat. Sci. Technol.*, **51**, 317–325.
Fersi C. Gzara L. and Dhahbi M. (2005). Treatment of textile effluents by membrane technologies. *Desalination*, **185**, 399–409.
Frechen F.-B., Schier W., Wett M. andWaldhoff A. (2006). Upgrading of existing municipal wastewater treatment plants with the 'non-conventional low MLSS MBR technology'. *Eng. Life Sci.*, **6**, 68–73.
Galil N. I. and Levinsky Y. (2007). Sustainable reclamation and reuse of industrial wastewater including membrane bioreactor technologies: case studies. *Desalination*, **202**, 411–417.
Gander M., Jefferson B. and Judd S. (2000). Aerobic MBRs for domestic wastewater treatment: a review with cost considerations. *Sep. Purif. Technol.*, **18**, 119–130.
González S., Petrovic M. and Barceló D. (2007). Removal of a broad range of surfactants from municipal wastewater – comparison between membrane bioreactor and conventional activated sludge treatment. *Chemosphere*, **67**, 335–343.
Harussi Y., Rom D., Galil N. and Semiat R. (2001). Evaluation of Membrane Processes to Reduce the Salinity of Reclaimed Wastewater. *Desalination*, **137**, 71–89.
Hassaine-Sadi F. and Sadoun L. (2005). Treatment of industrial wastes containing toxic metals (chromium). Purification by liquid membranes. *Desalination*, **185**, 335–340.
Khamis M., Manassra A., Dakiky M. (2005). *Proceedings of the International Conference on Water Values and Rights*, May 2–4, Ramallah, Palestine, pp. 344–355.
Khemekhem S., Ben Amar R., Ben Hassen R., Larbot A., Ben Salah A., and Cot L. (2006). Fabrication of mineral supports of membranes for microfiltration/ultrafiltration from tunisian clay. *Ann. Chim.*, **31**, 169–181.
Khider K. Akretche D. E. and Larbot A. (2004). Purification of water effluent from a milk factory by ultrafiltration using algerian clay support. *Desalination*, **167**, 147–151.
Kim D. S., Kang J. S. and Lee Y. M. (2004). Removal of dye from wastewater using micellar-enhanced ultrafiltration and recovery of surfactant. *Sep. Sci. Technol.*, **39**, 833–842.
Le-Clech P., Chen V. and Fane A. G. (2006). Fouling in membrane bioreactors used in wastewater treatment. *J. Membr. Sci.*, **284**, 17–53.
Lesjean B. and Luc F. (2006). Assessment of the membrane bioreactor technology and European market outlook. http://www.idswater.com/Common/Paper/Paper_155/ (accessed 12 May 2007).
Lesjean B., Rosenberger S., Laabs C., Jekel M., Gnirss R. and Amy G. (2005). Correlation between membrane fouling and soluble/colloidal organic substances in membrane bioreactors for municipal wastewater treatment. *Water Sci. Technol.*, **51**, 1–8.
Magiera P., Suzan T. Z. and Nolte L. (2006). Water demand management in the Middle East and North Africa. *Manag. Env. Qual.: An Intl. J.*, **17**, 289–298.
Mandil M. A. and Bushnak A. A. (2002). Future needs for desalination in South Mediterranean countries. *Desalination*, **152**, 15–18.
Masmoudi S., Ben Amar R., Feki H., Larbot A., Cot L. and Ben Salah A. (2006). Use of ultrafiltration membranes with apatite for the treatment of cuttlefish effluent. *Desalination*, **200**, 335–336.
Nghiem L. D., Manis A., Soldenhoff K. and Schäfer A. I. (2004). Estrogenic hormone removal from wastewater using NF/RO membrane. *J. Membr. Sci.*, 242 (2004) 37–45.
Oron G., Bick A., Gillerman L. and Manor Y. (2004). Hybrid membrane systems for secondary effluent polishing for unrestricted reuse for agricultural irrigation. *Water Sci. Technol.*, **50**, 305–312.
Oron G., Gillermana L., Buriakovskya N., Bickd A., Gargirb M., Dolanb Y., Manore Y., Katze L. and Haginc J. (2008). Membrane technology for advanced wastewater reclamation for sustainable agriculture production. *Desalination*, **218**, 170–180.
Ouazzani K. and Bentama J. (2008). Bio-fouling in membrane processes: micro-organism/surface interactions, hydrodynamic detachment method. *Desalination*, **220**, 290–294.
Quteishat K., Abu Arabi M. K. and Reddy K. V. (2003). Review of MEDRC R&D projects. *Desalination*, **156**, 1–20.
Rachwall T. and Judd S. (2006). A synopsis of membrane technologies in UK municipal potable water treatment: history, status and prospects. *Water Environ. J.*, **20**, 110–113.
Rebhun M. (2004). Desalination of reclaimed wastewater to prevent salinization of soils and groundwater. *Desalination*, **160**, 143–149.
Rosenberger S., Laabs C., Lesjean B., Gnirss R., Amy G., Jekel M. and Schrotter J.-C. (2006). Impact of colloidal and soluble organic material on membrane performance in membrane bioreactors for municipal wastewater treatment. *Water Res.*, **40**, 710–720.

Saddoud A., Ellouze M., Dhouib A., and Sayadi S. (2007). Anaerobic membrane bioreactor treatment of domestic wastewater in Tunisia. *Desalination*, **207**, 205–215.

Saffaj N., Alami-Younssi S., Albizane A., Messouadi A., Bouhria M., Persin M. and Larbot A. (2004). Preparation and characterization of ultrafiltration membranes for toxic removal from wastewater. *Desalination*, **168**, 259–263.

Semiat R. (2000). Desalination: present and future. *Water Int.*, **25**, 54–65.

Wang S., Guillen G. and van Hoek E. (2005). Direct observation of microbial adhesion to membranes. *Environ. Sci. Technol.*, 39, 6461–6469.

Wilderer P. A. (2005). Water supply and sanitation – a major challenge for Europe. *Water Sci. Technol.*, **51**, 1–6.

Yang W., Cicek N. and Ilg J. (2006). State-of-the-art of membrane bioreactors: Worldwide research and commercial applications in North America. *J. Membr. Sci.*, **270**, 201–211.

Younos T. J. (2005). Environmental issues of desalination. *Contemp. Water Res. Edu.*, **132**, 11–18.

Zhang J., Chua H. C. Zhou J. and Fane A. G. (2006). Factors affecting the membrane performance in submerged membrane bioreactors. *J. Membr. Sci.*, **284**, 54–66.

Chapter 3

Guideline of action regarding future research and development in the field of MT

Alessandra Criscuoli, Alberto Figoli, Rashed Al-Sa`ed and Enrico Drioli

INTRODUCTION

Water scarcity represents today a problem of big concern, not only for Arabian or African countries, but for the whole earth. The increase of population, as well as the needs of modern society which are leading to a higher development of industrial productions (that means a higher use of water as feed source and a higher production of wastewater), are at the basis of the increase of the fresh water demand registered in last years. In this context, membrane technologies are playing an important role. Wastewater treatment by membrane operations is an interesting way to recover purified water from the effluent streams coming from industries, by improving the overall performance of the production process. This is the case, for example, of textile industries that discharge aqueous solutions containing dyes to the environment. By treating these streams by ultrafiltration, nanofiltration, reverse osmosis, membrane distillation, it is possible to obtain a water stream without dyes and a stream concentrated in dyes that can be recycled to the process (Calabrò *et al.*, 1990; Ciardelli *et al.*, 2000; Marcucci *et al.*, 2002; Tang and Chen, 2002; Purkait *et al.*, 2004; Mozia *et al.*, 2005; Mozia *et al.*, 2007; Van der Bruggen *et al.*, 2004; Banat *et al.*, 2005; Criscuoli *et al.*, 2008). In this way, the 'sustainability' of the process is ensured. Other interesting cases are the recovery of heavy metals from water by extraction by membrane contactors or by using supported liquid membranes (Yun *et al.*, 1993; Ho *et al.*, 2000; Ho and Poddar, 2000; Lin and Juang, 2001), the volatile organic compounds (VOCs) removal by pervaporation (Peng *et al.*, 2003) and the arsenic removal by nanofiltration or reverse osmosis (Waypa *et al.*, 1997; Brandhuber and Amy, 1998; Urase *et al.*, 1998; Kang *et al.*, 2000; Ning, 2002; Vrijenhoek and Waypa, 2000; Saitua *et al.*, 2005). More recently, also the arsenic oxidation (from As(III) to As(V)) with air by membrane contactors, has been successfully demonstrated (Galizia, 2007). Other examples refer to the treatment of olive oil wastewater streams (Molinari and Drioli, 1988; Borsani and Ferrando, 1996; Paraskeva *et al.*, 2007; Russo, 2007) or of aqueous streams coming from tannery (Cassano *et al.*, 1999). Membrane bioreactors (MBRs) combine membrane filtration with biological treatment and are effective for treating industrial wastewater, domestic wastewater and specific municipal wastewater, where a small footprint, water reuse, or stringent discharge standards are required. Promising applications also exist in treating landfill leachate, chlorinated solvents in manufacturing wastewaters, wastewater polluted with trace pharmaceutical compounds, for groundwater remediation, and for the treatment of wastewaters from petrochemical industry (Qin *et al.*, 2007). MBRs are considered by the European Union as one of the Best Available Technologies (BAT) for municipal and industrial wastewater treatment and the increased interest in the MBR technology is clearly evident from the high numbers of projects of research funded (http://eippcb.jrc.es). Compactness, reduced sludge production, high efficiency in physical disinfection are the main advantages claimed for these systems (Qin *et al.*, 2007).

An intensive application of membrane operations is also in the water desalination field. Today, reverse osmosis is the dominant technology and all new installed plants use reverse osmosis units for desalting water. Moreover, as a pre-treatment step, microfiltration and ultrafiltration units are more often present, instead of more traditional coagulation or sand filtration devices (Glueckstern *et al.*, 2001).

The number of membrane-based desalination plants is, therefore, increasing worldwide, accounting for about 80% of all desalination facilities (Frenkel 2008). Larger plants were building and also bigger RO modules were designed and developed (16 inch modules with membrane areas of 1570–1725ft^2 instead of the traditional 8 inch modules with membrane areas of 365–400ft^2), both to reduce operating and capital costs (because of the capability to operate at higher flow rates) and to work with higher capacities (Industry News, 2008; Hallan *et al.*, 2008).

The investments in RO technology are increasing, as well as the number of projects on seawater desalination. Global desalination capacity is expected to reach 107million m^3/d in 2016 and it has been calculated that over the decade 2007–2016 $64billion will be spent on building desalination plants (Industry News, 2008). This scenario states the extreme importance for Mediterranean countries to 'take part' to the evolutions in the field and, therefore, to overcome the existing gap with more-advanced countries.

3.1 POINTS OF ACTION FOR IMPROVING THE PERFORMANCE OF MEMBRANE TECHNOLOGIES IN WATER TREATMENT

Although membrane operations are largely used in water treatment, the continuous need of increasing the fresh water recovery factor and the water quality, while reducing the water cost and the environmental impact, has led to the research of methods for improving the performance of membrane units (United Nations, 2001; Republic of South Africa, 2005/2006). As a matter of fact, different are the drawbacks still present that came out from the application 'on field' of membrane technologies (see Table 3.1). For example, the fouling and scaling issues have to be solved in order to improve the performance of the system, by ensuring high recovery in long-term operations. Improvements in the pre-treatment steps will be crucial in this sense and also for the reduction of overall costs (pre-treatment can account up to 60% of the whole costs in seawater desalination). In last years, microfiltration and ultrafiltration have been introduced as pre-treatment steps, instead of traditional devices such as sand filtration and flocculation. The reduction of chemical usage is useful from both an economical and environmental point of view.

Table 3.1 Main drawbacks of membrane-based water treatment plants

Fouling and scaling
Corrosion
Concentration polarization
Low membrane resistance to chemicals and high temperatures
Low membrane productivity at low feed pressures
Low MBR efficiency in medium-high saline environments
Chemicals usage and effluents treatments
Brine disposal
Energy consumption
Water cost

Among the different types of fouling, bio-fouling can have several adverse effects on RO systems (Kramer and Tracey, 1995), such as membrane flux decline, membrane biodegradation, increased salt passage, increased differential pressure and feed pressure, as results from the formation of bio-film resistance.

Recently, MBRs have gained success in the treatment of wastewaters and studies on their application in the pre-treatment phase for the abatement of the organic compounds present on brackish and seawater (biofouling reduction) are in progress. As already reported, MBRs have the advantages of small foot print, reduced sludge production and high quality of effluent from a sanitary point of view (Qin *et al.*, 2007). The research activities on these units are mainly focused on the development of effective methods to limiting membrane clogging and operational failures, on the reduction of costs for maintenance and replacement of membranes, on keeping efficiency even in medium-high saline environments. A proper training of the committed staff is also essential for ensuring fast actions in case of any process or system disturbance. A help for the right choice of the pre-treatment steps comes from the membrane autopsy. It allows, in fact, defining the compounds that more affected the membrane performance during its life in the plant and also the type of damage created.

Module configuration and design is also important for a reduction of fouling, as well as of the concentration polarization, and for an increase of the water recovery factor. The reduction of the mass transport resistances and pressure drops inside the module highly affects the efficiency of the process. Modelling studies are quite useful for the design of both the module and the membrane-self. Concerning the latter, transport models for the membrane taking into account the effect of pore size, membrane material, surface electrical charge, surface roughness, etc. allow to predict the membrane performance under a variety of conditions, already before its installation, thus, addressing the membrane manufacturers.

The development of new membranes is, in fact, at the basis of the research efforts devoted to the improvements of water treatment plants. Novel biofouling resistant membranes, for example, are of big interest for reducing fouling issues and chemical cleanings. Membranes with higher recovery factors will lead to a reduction of the use of raw water and of the specific energy consumption. Membranes able to withstand high operating pressures and temperatures (higher life-time), with higher salt rejections and chemical resistance, or able to efficiently work with lower feed pressures (lower energy consumption), will positively affect the overall performance of the plant, reducing the overall costs. Operating costs can be reduced through the development of systems of automation and control that will help in simplifying the management of the plant.

The energy cost represents one of the main components of the costs associated to a membrane-based water treatment plant and, therefore, different strategies for reducing the energy consumption are under investigation. As already mentioned, the development of high recovery factor membranes able to work with lower feed pressures and the improvement of pre-treatment steps are possible means to decrease the energy costs. Energy saving can be achieved also by installing energy recovery devices and/or by using renewable energies. There are several research projects in progress on desalination driven by solar thermal energy, where solar thermal collectors are used in the desalination plant (http://www.pas/es/webeng/projects/medesol/index.html). Other possible alternative sources of energy are the thermal energy delivered by a salinity-gradient solar pond and wind power systems.

The environmental impact of a water treatment plant is another point of big concern. Special attention must be given to the chemicals discharge and, in desalination, to the brine disposal, not only in terms of effluent quality, but also by properly choosing the location of the effluent outfall. The development of procedures (such as Life Cycle Analysis) for assessing the environmental impact of the desalination plant effluents can be helpful for better addressing the research on this matter. Chemicals usage in the plant has to be reduced as much as possible, in order to match the legislation requirements for effluent streams. This means to operate with higher efficient systems (lower fouling and, therefore, reduction of chemical cleaning cycles), to improve the post-treatment section, or to introduce innovative systems into the plant. For example, in desalination plants, besides fouling, corrosion phenomena can occur, causing the release of heavy metals into the aqueous streams and an increase of maintenance costs. The addition of chemicals and the deareation in packed towers are usually employed for limiting this problem. Chemicals are also used for controlling the pH and the conductivity of water. New membrane systems, membrane contactors (Drioli *et al.*, 2006a), have recently been applied for controlling the oxygen and carbon dioxide content into water, without the use of chemicals (http://medina.unical.it). In these systems, the liquid stream is blocked at one side of microporous hydrophobic membranes (in correspondence of the pore mouths), while a gas phase is sent at the other side of the membranes. Due to a difference in partial pressure across the membranes, there is a transfer of the gaseous species from/to the liquid stream. Figure 3.1 shows this system for the simultaneous removal of oxygen and the addition of carbon dioxide.

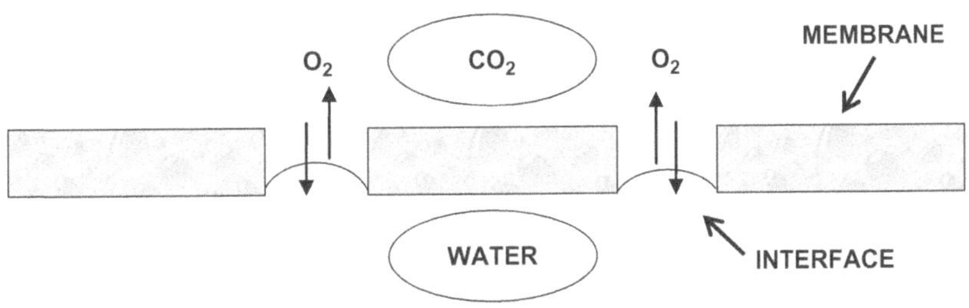

Figure 3.1 Simultaneous transfer of oxygen and carbon dioxide in a membrane contactor

Referring to the brine disposal, a new way to try to solve this problem is the use of membrane crystallizers working on the reverse osmosis or nanofiltration concentrates. Researches on this topic are in progress and the preliminary results obtained are quite encouraging (Drioli *et al.*, 2004). The membrane crystallizer consists of a membrane distillation device pushed up to the crystals formation and the outputs of the system are fresh water and solid crystals (see Figure 3.2). This technique will, in principle, be able to lead to a zero brine discharge, with a significant improvement in the environmental impact.

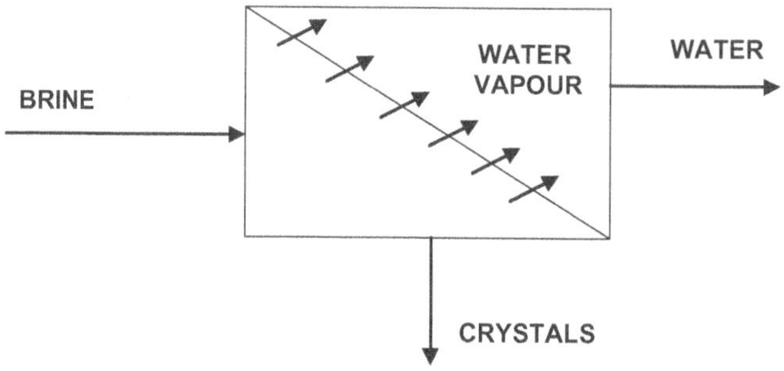

Figure 3.2 Production of pure water and solid crystals in a membrane crystallizer

The strategy of re-designing desalination plants by integrating different membrane units is becoming quite interesting for increasing the water recovery factors of the existing plants, while reducing the chemical consumptions and the fouling and environmental issues (Drioli *et al.*, 2002; Drioli *et al.*, 2006b). Figure 3.3 shows a typical integrated membrane system proposed for desalination (http://medina.unical.it). The pre-treatment step can be made of microfiltration and ultrafiltration (MF/UF) units or of membrane bioreactors (MBR). Nanofiltration (NF) can work before reverse osmosis (RO), to further reduce the fouling and to increase the water recovery factor. The membrane contactor (MC) unit has the role to control the gas composition of the streams and can be used on the nanofiltration and reverse osmosis permeates or brines. Nanofiltration and reverse osmosis brines can be treated by a membrane crystallizer (MCr), to produce salt crystals and other fresh water.

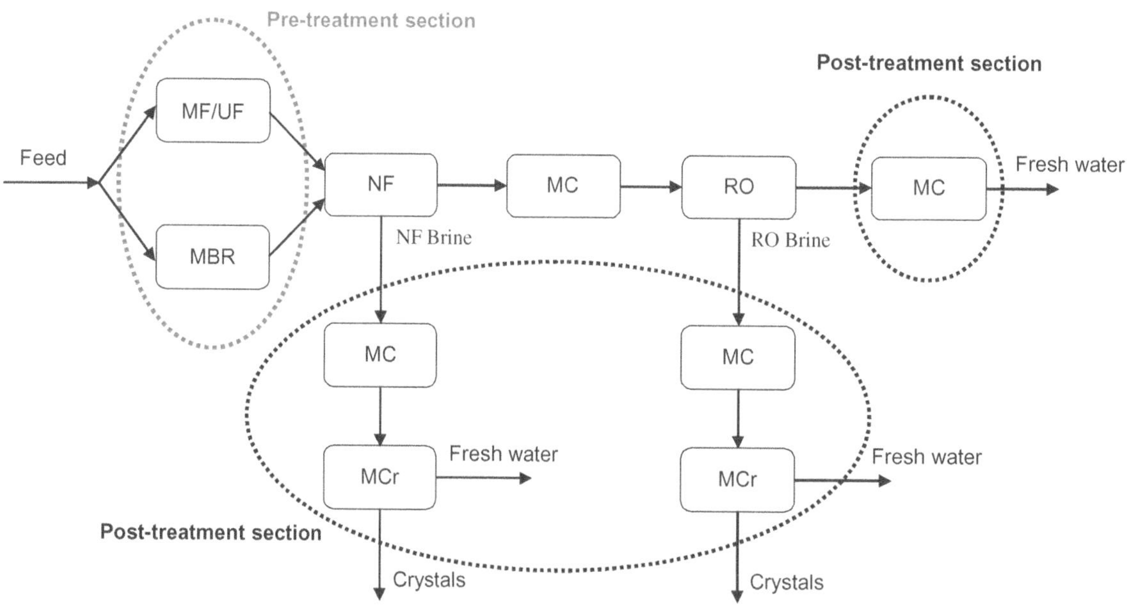

Figure 3.3 Integrated membrane system for water desalination

The water treatment plants must take into consideration also the health impact of the produced water on our daily life. In this sense, microbial or chemical contamination must be carefully checked and both pre-treatment and post-treatment steps have to be designed to improve the water quality and to avoid any possible contamination of the final product. This aspect becomes more important if we think that high-quality water sources are reducing and that lower-quality sources have to be considered as intake of the treatment plants. The choice and management of water resources is another point of concern. It should be possible to guarantee both the environmental protection and an efficient utilization of the resources. Social, environmental and economical factors must, therefore, be taken into account.

When proposing a new technology or new design concepts, it is also important to consider the cost, the simplicity of the new systems in terms of operation and maintenance, as well as the availability of local infrastructures for their installation. Often, poor countries are those more affected by the problem of water scarcity and also auxiliary capabilities are missing. Cost-effective technologies not requiring a significant amount of infrastructures must be developed, as well as appropriate training and educational stages about the system and water management.

Based on the considerations made, it seems clear that the research in the field must take into account different factors in order to be defined 'sustainable' and meet the requirements of future generations. Table 3.2 summarizes the main points to focus on for further developing the existing technologies.

Table 3.2 Research orientations

Process and membrane modelling and design
Analysis of fouling, scaling, corrosion phenomena and methods for their control
Development of membranes with higher productivity and life-time
Development of membranes with higher resistance to biofouling, chemicals and high temperatures
Development of membranes able to work with lower pressure feeds
Development of modules with reduced pressure drops and mass transport resistances
Development of new pre-treatment steps for reducing fouling and scaling issues
Application of MBRs in medium-high saline environments
Use of energy recovery devices and/or renewable energies for energy saving
Development of procedures (such as Life Cycle Analysis) for assessing the environmental impact of the desalination plant effluents
Reduction of chemicals usage by introducing new membrane systems in the plant
Design and development of integrated membrane systems for improving the fresh water recovery, the water quality, and reducing environmental impact
Efficient management of water resources while protecting the environment
Appropriate selection of sites for intake and outfalls
Reduction of the cost of the existing plants
Development of cost-effective and easy-to handle technologies
Automation and control of the plants
Share of knowledge by means of training and educational stages

3.2 IDENTIFICATION OF 'DEVELOPMENT ROUTES' FOR MEDITERRANEAN REGIONS

The existing gap in technology between Mediterranean regions and the 'more-advanced' countries is due to different factors and, among them, the poverty and the governmental situation are playing an important role. The governments are responsible of the development of the regions that can be obtained by investing in research, both at academic and industrial level and promoting the presence of companies and linkages between academia and industry. The poverty of the regions can sometimes limit these types of actions and obliges to choice low-cost and easy-to-manage technologies that not always represent the optimal solutions, both in terms of environmental impact and overall efficiency. In this scenario, investments in the sharing of knowledge become a possible answer to the need of development of these countries. As reported in the section above, the research activities carried worldwide on water treatment plants have evidenced several 'point of weakness' of the existing technologies, as well as 'points of actions' to follow in order to overcome the identified drawbacks. Therefore, the sharing with Mediterranean Countries of the acquired knowledge would be a good starting point. Specifically, research institutions and academia, industries and operators working in the field, have to be involved in the sharing. It is, in fact, of fundamental importance to promote training and dissemination of knowledge, in order to educate the 'users' in the more update and complete way. How to do this in practice? One option is the creation of research clusters. The potential clusters in the Mediterranean and North Africa regions have been identified within the ProMembrane project and are illustrated in Figure 3.4 (Al-Sa`ed *et al.*, 2009).

Besides the interactions among these countries, the cooperation with the 'more advanced countries' must be established. This can be particularly obtained by:

- Creating joint projects, for example within those promoted by the European Commission.
- Engaging in bilateral cooperation with European countries.
- Fostering participation of students in study abroad programmes.

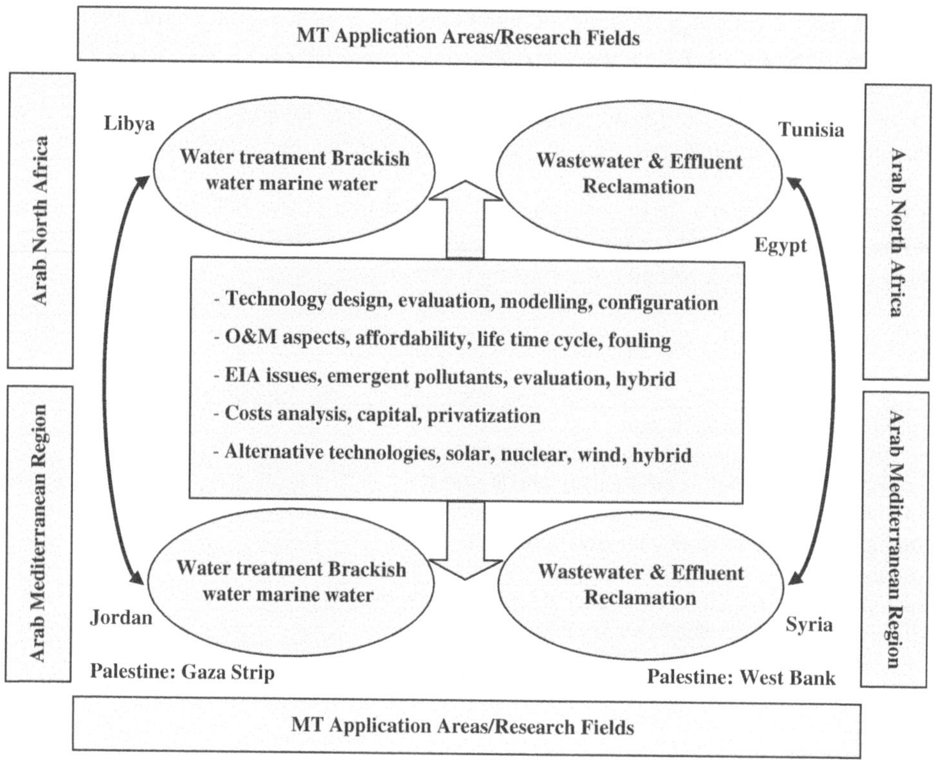

Figure 3.4 Research clusters identified in ProMembrane

3.2.1 Creating Joint Multilateral Projects

The European Union offers several possibilities for institutions from Mediterranean Partner Countries (MPCs) to engage in multilateral joint projects. Some examples are the Europe Aid Programme and the Framework Programmes for Research and Technological Development. Within these programmes, differing projects of cooperation between organisations from European Member States and from Asia or from MPCs have been approved and carried out. For example, the **INNOWA** project: 'Innovative water treatment' (http://www.innowa.org/), had as partners organisations from Italy, Germany, China and Bangladesh with the aim of building capacity for technology transfer by establishing a European-Asian technology network on innovative water treatment technologies with focus on membrane techniques. The **INNOVA-MED** project: 'Innovative processes and practices for wastewater treatment and re-use in the Mediterranean region' could also be a good example (http://www.cid.csic.es/innova-med/home.htm), as well as **PROMEDACCESS** 'Promotion of the participation of Researchers from Mediterranean Partner Countries to European Research and Mobility Programmes' (http://www.inp.eg.net/PROMEDAccess.html). PROMEMBRANE (www.promembrane.info) itself is a good example of project to consider for this scope. More international joint projects can be found at the ProMembrane Public Library, in the Database of Membrane Technology Experts, as well in the CORDIS database of projects financed by the FP6 (http://cordis.europa.eu/fp6/projects.htm).

The FP6 (Sixth Framework Programme for Research and Technological Development) time frame is over and even though some projects are still running, participation in future projects is now possible through the FP7, in which there are different calls that would be of interest for promoting international cooperation. Particularly, it is the possibility of 'Integration of International Cooperation in all Specific Programmes'. Figure 3.5 shows the FP7 structure.

The new emphasis on international research cooperation of FP7 has the objectives of:

- Improving EU **scientific and economic development** by means of **strategic partnerships** with Third Countries;
- Ensuring **better access** for Third Countries to research carried out elsewhere in the World;
- Facing **specific problems** of Third Countries.

In the different areas of FP7, the international cooperation can be in terms of:

- **Cooperation**: collaborative projects between different global research partners and European researchers;
- **Ideas**: excellent projects encouraging top international researchers to participate in Europe-led teams;
- **People**: incoming and outgoing international mobility of researchers (Marie Curie Actions);
- **Capacities**: strengthen research capacity in Europe and other regions of the world.

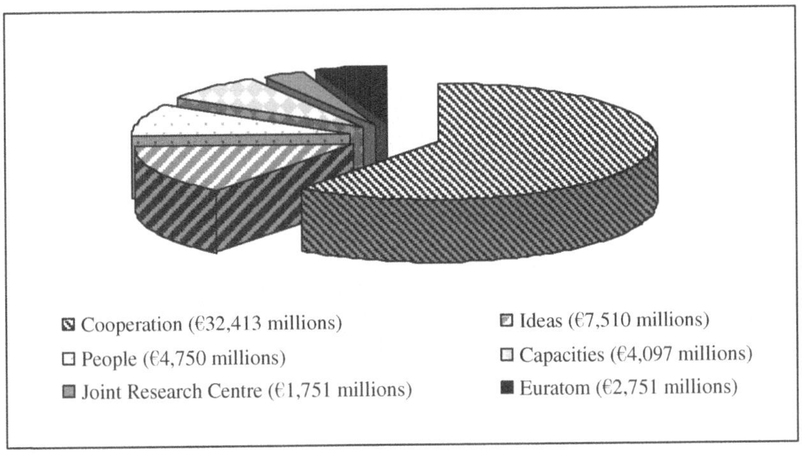

Figure 3.5 FP7 structure and budget distribution

The funding schemes of the FP7 are summarized in Figure 3.6, and the main aspects of the international cooperation within the different areas are briefly described in Figures 3.7–3.10 (http://www.cordis.lu).

Collaborative Projects (Small/medium and large)

Networks of Excellence

Coordination and Support Actions

Individuals projects

Support for training and career development of researchers

Research for the benefit of specific groups-in particular SMEs

Figure 3.6 Funding schemes in FP7

All themes are opened to Third Countries

- **ICPC** funded in all cases
- Industrialised countries only if indispensable
- Participation of Third Countries is encouraged
- 3 partners from MS or AC; additional partners from Third Countries are optional

Specific **I**nternational **C**ooperation **A**ctions (**SICA**) in specific topics of mutual interest

- Dedicated calls or separate dedicated parts of calls targeted to ICPC countries
- At least four partners: two Member or Associated States and two ICPC

Figure 3.7 International cooperation in COOPERATION

ERC specific programme encourages participation of **excellent researchers** from Third Countries

Aims:

- to integrate EU into the worldwide research community
- to open to Third Countries on basis of mutual benefit or according to relevant concluded agreements
- participation of Third Countries to be justified on basis of added scientific value

Figure 3.8 International cooperation in IDEAS

Aim to reinforce European research with foreign scientists

To build lasting connections through mobile researchers

Two main actions:

- Career development/life long training for EU researchers
  - **International outgoing fellowships**
  - **International re-integration fellowships**
- International cooperation for and with researchers from Third Countries
  - **International incoming fellowships for experienced researchers**
  - **Marie Curie host driven actions**

Figure 3.9 International cooperation in PEOPLE

Overall coordination of international cooperation

Aims:

- to identify **S&T priorities** with Third Countries
- to support and stimulate their **participation** in FP7
- to reinforce bilateral S&T **cooperation** with targeted Countries
- to coordinate S&T **national programs** with Third Countries

Three main activities:
Bi-regional coordination of S&T Cooperation "INCO-NET"
Bi-lateral S&T cooperation partnerships "BILAT"
Coordination of national policies and activities "ERA-NET"

Figure 3.10 Other opportunities to interact within the FP7

Figure 3.11 shows types of participants to FP7. Mediterranean Partner Countries are clearly reported and they are: **Algeria, Egypt, Jordan, Lebanon, Libya, Morocco, Palestine, Syria, and Tunisia**. Besides the international cooperation, there are other opportunities to interact, as reported in Figure 3.12.

Types of participants

- Research groups at Universities or Research Centres
- SMEs (small to medium-sized enterprises)
- Public Administrations, Multinational corporations
- individuals, from early-stage to experienced researchers

Countries

- EU-27
- Candidate Countries and Associated States
- **International Cooperation Partner Countries (ICPC)**
  - Western Balkans countries
  - **Mediterranean Partner countries**
  - Eastern European and Central Asian countries
  - Africa Caribbean Pacific
  - Latin America
  - Asia

Figure 3.11 Types of participants to FP7

The **European Neighbourhood and Partnership Instrument (ENPI)**
EC Financial Instrument to Eastern Europe, Southern Caucasus and South Mediterranean countries since 1st January 2007. It replaces the MEDA and TACIS instruments.

Eligible Countries: Algeria, Armenia, Azerbaijan, Belarus, Egypt, Georgia, Israel, Jordan, Lebanon, Libya, Moldova, Morocco, the Palestinian Authority, Syria, Tunisia and Ukraine.

Budget: A total of over EUR 11.1 billion for the 2007-13 period

Types of programmes:

- National programmes for each partner country
- Three Regional programmes: one for the East, one for the South and one trans-regional for both East and South.
- Fifteen Cross-border Cooperation programmes.
- Five thematic programmes. They are common to neighbouring countries and development countries (ACP, ALA..)

Figure 3.12 Other Opportunities to interact within the FP7

The European Community Research & Development Information Service (**CORDIS**) offers more details and useful information for those who are interested in FP7 actions. For example, at http://cordis.europa.eu/en/home.html it is possible to find calls, find partners, as well as to get support. Specifically, through the 'Get support' link, FP7 contacts in third countries can be obtained. This last point is quite important providing support to the organisations from Mediterranean and North Africa regions. Moreover, at https://www.epss-fp7.org/epss/EPSS-Userguide.pdf an Electronic Proposal Submission Service (EPSS) preparation and submission guide can be downloaded to know exactly what steps follow for participating to a project proposal within a certain call (the guide is provided as Appendix to this deliverable). Some other useful links are:

- Seventh Framework Programme (FP7): http://ec.europa.eu/research/fp7/index_en.cfm
- Information on research programmes and projects: http://cordis.europa.eu/fp7/home_en.html
- International Cooperation: http://ec.europa.eu/research/iscp/index.cfm

3.2.2 Bilateral Co-Operations with European Countries

Bilateral projects can be another mean for creating international cooperations. In Italy, both the National Council of Research (CNR) and the Minister of Foreign Affairs are promoting these types of actions between Italy and most of the Mediterranean Countries, such as Tunisia, Algeria, Morocco, Egypt, etc. Specifically, bilateral projects are now in progress between ITM and Middle East (MEDRC) and Egypt. In Germany, the Federal Ministry of Education and Research (BMBF) fosters bi- and multilateral cooperation in education and research with countries from the Middle East (Jordan) and Africa (Egypt and South Africa). For instance, a bilateral initiative of the German and the Egyptian Ministries of Higher Education and Research is the German-Egyptian Year of Science and Technology, where educational and technological institutions from both countries are given the chance to promote a broader scientific exchange and research cooperation. Besides, the German Academic Exchange Service (DAAD) as well as the Humboldt Foundation also offers the possibility to engage in cooperation with institutions from countries of the Middle East and North Africa.

The strengthening of educational programmes in MPC should also be an aim of the international cooperation among countries. Therefore, another recommendation would be the development of joint projects for the improvement of higher education programmes (e.g. the DAAD recently published a call for proposals for the conception and implementation of an international master course ‘Renewable Energies (Energy Efficiency)’ for German and Egyptian specialists in Egypt).

Within this scope, the Italian National Council of Research (CNR) has undertaken the ‘Mediterranean Programme’ to foster the development of human resources, rapprochement between cultures, and cooperation in social and economic reforms in the Third Mediterranean Countries. Appropriate high-level training activities are identified and coordinated within the programme.

3.2.3 Fostering the Participation of Students in Study Abroad Programmes

Students should engage in programmes to study abroad. According to the Results of the 17th Social Survey of the Deutsches Studentenwerk (DSW) conducted by HIS Hochschule-Information-System, most foreign students at German institutions of higher education still come from European countries. Of the 227,000 foreign students enrolled in the winter semester 2002/03, more than 136,000 were Europeans, whereas only 11,800 were from North Africa and 12,906 from the Middle East.

3.3 CONCLUDING REMARKS

Based on the considerations made, future research on membrane operations for water treatment must be devoted to the development of technologies able to be highly efficient and in the meantime sustainable. Efforts for reducing energy, chemicals and raw material consumptions and costs of the existing plants are, for example, required. Moreover, membranes with higher productivity, higher resistance to biofouling, chemicals and high temperatures, and higher life-time are needed, together with modules characterized by reduced pressure drops and mass transport resistances.

Referring to the creation and sharing of knowledge, different are the possibilities available. Recommendation for Mediterranean Partner Countries is therefore to be active in these fields, applying for joint projects, especially for those devoted to the training and educational on new technologies. In this way, it will be possible to improve their expertise and minimise the existing gap.

3.4 REFERENCES

Al-Sa`ed R., Sayadi S., Ghata A., Abdel-Shafy H., Schories G., Oropeza M., Lorenzo A. and Drioli E. (2009). Advancing membrane technologies for wastewater treatment and reclamation in selected Arab MENA countries. *Desalin. Water Treat.*, **4**, 287–293.

Banat F., Al-Asheh S. and Qtaishat M. (2005). Treatment of waters colored with methylene blue dye by vacuum membrane distillation. *Desalination*, **174**, 87–96.

Borsani R. and Ferrando B. (1996). Ultrafiltration plant for olive vegetation waters by polymeric membrane batteries. *Desalination*, **108**, 281–286.

Brandhuber P. and Amy G. (1998). Alternative methods for membrane filtration of arsenic from drinking water. *Desalination*, **117**, 1–10.

Calabrò V., Pantano G., Kang M., Molinari R. and Drioli E. (1990). Experimental study on integrated membrane processes in the treatment of textile effluents – energy and exergy analysis. *Desalination*, **78**, 257–277.

Cassano A., Criscuoli A., Drioli E. and Molinari R. (1999). Clean operations in the tanning industry: aqueous degreasing coupled to ultrafiltration. Experimental and theoretical analysis. *Clean Prod. Process.*, **1**, 257–263.

Ciardelli G., Corsi L. and Marcucci M. (2001). Membrane separation for wastewater reuse in the textile industry. *Resources, Conservation and Recycling*, **31**, 189–197.

Criscuoli A., Zhong J., Figoli A., Carnevale M. C., Huang R. and Drioli E. (2008). Treatment of dye solutions by vacuum membrane distillation. *Water Res.*, **42**, 5031–5037.

Drioli E., Criscuoli A. and Curcio E. (2002). Integrated membrane operations for seawater desalination. *Desalination*, **147**, 77–81.

Drioli E., Criscuoli A. and Curcio E. (2006a). Membrane contactors: fundamentals, applications and potentialities. Elsevier, Amsterdam.

Drioli E., Curcio E., Criscuoli A. and Di Profio G. J. (2004). Integrated system for recovery of CaCO3, NaCl and MgSO4·7H2O from nanofiltration retentate. *Journal of Membrane Science*, **239**, 27–38.

Drioli E., Curcio E., Di Profio G., Macedonio F. and Criscuoli A. (2006b). Integrating membrane contactors technology and pressure-driven membrane operations for seawater desalination: energy, exergy and costs analysis. *Chemical. Engineering Research and Desalination., Trans IChemE, Part A.* **84** (A3), 209–220.

Frenkel V. (2008). Brackish vs seawater desalination: Which one is for you? *International Desalination and water reuse.*, **17**(4), 47–50.

Galizia A. (2007). Treatment of water containing arsenic by means of membrane operations. Master thesis, University of Calabria, Italy.

Glueckstern P., Thoma A. and Priel M. (2001). The impact of R&D on new technologies, novel design concepts and advanced operating procedures on the cost of water desalination. *Desalination*, **139**, 217–228.

Hallan M. J., Johnson J. E., Koreltz M. S., Peery M. H. and Peterson G. W. (2008). Design, development and evaluation of 16-in diameter RO modules. *International Desalination and water reuse*, **17**(4), 20–27.

Ho W. S. W., Kalini S. and Poddar T. K. (2000). *Proceedings of the AIChE Spring National Meeting*, Atlanta, GA, USA, pp. 10–15.

Ho W. S. W. and Poddar T. K. (2000). *Proceedings of the AIChE Spring National Meeting*, Atlanta, GA, USA, pp. 38–43.

MEDINA: membrane-based desalination: an integrated approach (2006–2010). A research project supported by the European Commission under the Sixth Framework Programme (FP6). University of Calabria, Italy. http://medina.unical.it (accessed December 2007).

MEDESOL: seawater desalination by innovative solar-powered membrane-distillation system (2006–2010). A research project supported by the European Commission under the Sixth Framework Programme. Centro de Investigaciones Medioambientales y Tecnológicas, Spain. http://www.psa/es/webeng/projects/medesol/index.html.

Industry News (2008). *IDA Water News* 3.

Kang M., Kawasaki M., Tamada S., Kamei T., Magara Y. (2000). Effect of PH on the removal of arsenic and antimony using reverse osmosis membranes. *Desalination*, **131**, 293–298.

Kramer J. F. and Tracey D. A. (1995). *Proceedings of the IDA World Congress on Desalination and Water Reuse*, Abu Dhabi, Vol. IV, p. 3344.

Lin S.-H. and Juang R.-S. (2001). Mass-transfer in hollow-fiber modules for extraction and back-extraction of copper (II) with LIX64N carriers. *J. Membr. Sci.*, **188**, 251–262.

Marcucci M., Ciardelli G., Matteucci A., Ranieri L. and Russo M. (2002). Experimental campaigns on textile wastewater for reuse by means of different membrane processes. *Desalination*, **149**, 137–143.

Molinari R. and Drioli E. (1988). Processi integrati di ultrafiltrazione e osmosi inversa nel trattamento delle acque reflue da frantoi oleari. *Acqua-Aria*, **5**, 579–588.

Mozia S., Tomaszewska M. and Morawski A. W. (2005). A new photocatalytic membrane reactor (PMR) for removal of azo-dye Acid Red 18 from water. *Appl. Catal. B: Environ.*, **59**, 133–139.

Mozia S., Tomaszewska M. and Morawski A. W. (2007). Photodegradation of azo dye Acid Red 18 in a quartz labyrinth flow reactor with immobilized TiO_2 bed. *Dyes Pigme.*, **75**, 60–66.

Ning Y. R. (2002). Arsenic removal by reverse osmosis. *Desalination*, **143**, 237–241.

Paraskeva C. A., Papadakis V. G., Tsarouchi E., Kanellopoulou D. G. and Koutsoukos P. G. (2007). Membrane processing for olive mill wastewater fractionation. *Desalination*, **213**, 218–229.

Peng M., Vane L. M. and Liu S. X. (2003). Recent advances in VOCs removal from water by pervaporation. *J. Hazard. Mater.*, **98**, 69–90.

Purkait M. K., DasGupta S. and De S. (2004). Removal of dye from wastewater using micellar enhanced ultrafiltration and regeneration of surfactant. *Sep. Purif. Technol.*, **37**, 81–92.

Qin J.-J. Oo M. H., Tao G. and Kekre K. A. (2007). Study on petrochemical wastewater treatment and reuse using submerged MBR. *J. Membr. Sci.*, **293**, 161–166.

Russo C. (2007). A new membrane process for the selective fractionation and total recovery of polyphenols, water and organic substances from vegetation waters (VW). *J. Membr. Sci.*, **288**, 239–246.

Saitua H., Campderros M., Cerutti S. and Padilla A. P. (2005). Effect of operating conditions in removal of arsenic from water by nanofiltration membrane. *Desalination*, **172**, 173–180.

Tang C. and Chen V. (2002). Nanofiltration of textile wastewater for water reuse. *Desalination*, **143**, 11–20.

United Nations (2001). *Water Desalination Technologies in the ESCWA Member Countries. Report.* United Nation, New York.

Urase T., OH J. and Yamamoto K. (1998). Effect of pH on rejection of different species of arsenic by nanofiltration. *Desalination*, **117**, 11–18.

Van der Bruggen B., Curcio E. and Drioli E. J. (2004). Process intensification in the textile industry: the role of membrane technology. *J. Environ. Manage.*, **73**, 267–284.

Vrijenhoek E. M. and Waypa J. (2000). Arsenic removal from drinking water by a "loose" nanofiltration membrane. *Desalination*, **130**, 265–277.

Waypa J., Elimelech M. and Hering J. G. J. (1997). *J. AWWA*, **89**, 102–114.

Yun C. H., Prasad R., Guha A. K. and Sirkar K. K. (1993). Hollow fiber solvent extraction removal of toxic heavy metals from aqueous waste streams. *Ind. Eng. Chem. Res.*, **32**, 1186–1195.

Chapter 4

Related initiatives

Antonia Lorenzo and Aníbal Vega

4.1 JOINT RESEARCH ACTIVITIES

4.1.1 Projects funded under the EC Framework Programmes

In the recent years, the European Commission has supported through diverse financial instruments, a great number of membrane-based research projects. One of these instruments are the Framework Programmes (FP), which have shown a strong increment on membrane related initiatives funded between the 5th and 6th FP, comprising the period from 1998 to 2007.

Looking at the information on membrane related projects funded by the FPs provided by the databases of the EC website, http://cordis.europa, it can be shown that during the period 1998–2007, 22 membrane related projects were funded from which 7 were held during the 5th and the other 15 along the 6th Framework Programme.

As shown in Figure 4.1, the European Commission, concerned about the relevance of membrane technology, incremented 353% the funding provided to membrane related projects during the 6th FP in comparison with the amount invested during the 5th FP, and this trend is likely to continue during the 7th FP which is currently running (2007–2013). Projects like RSMBR, MEMBRIDGE, MEDIRAS, NEW-ED, IMETI and NAMETECH have been already funded with 7.5 M€. Therefore, it is clear the importance of providing economic resources to a technology that potentially can solve some of the water problems which affect many countries.

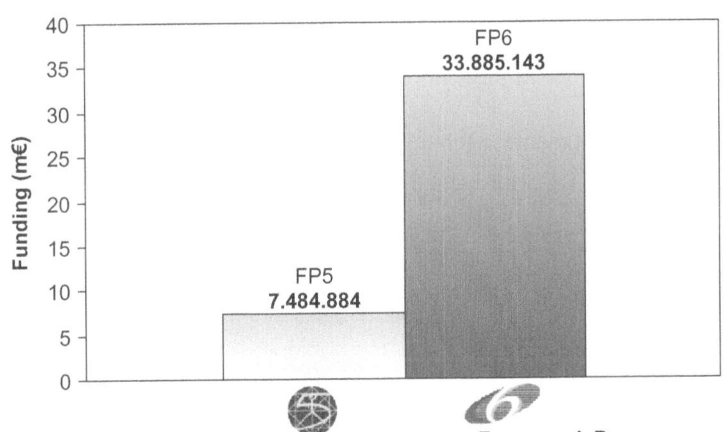

Figure 4.1 Evolution of funding on membrane related projects

Table 4.1 shows the 22 mentioned projects, given their title and acronym, amount of funding given by the EC, duration and the date of completion.

Table 4.1 Research projects membrane related funded within the 6th and the 7th FP

FP	Project	Acronym	Funding (€)	Duration (months)	Completion
FP 5	Transnational project for membrane bioreactor for textile plants	Transbiotex	565.936	24	November 1998
	O2 enrichment for bio-membrane reactors for improving treatment efficiency and capacity of new and existing waste water treatment systems	Itewat	231.404	12	May 2002
	Water recycling and reuse by application of Membrane BioReactors: textile and municipal wastewater as examples	MBR-Reclycling	540.000	39	May 2003
	Assessment of technologies for the removal of pharmaceuticals and personal care products in sewage and drinking water facilities to improve the indirect potable water reuse	Poseidon	1.372.963	42	June 2004
	Removal of persistent polar pollutants through improved treatment of wastewater effluents	P-Three	1.409.198	39	December 2005
	Integrated concepts for reuse of upgraded wastewater	Aquarec	2.299.521	36	February 2006
	Waste water recycling in textile finishing through the application and further development of membrane bio-reactors used in space life, support systems	Space2Tex	1.066.132	108	December 2011
FP 6	Innovative wastewater treatment applications for isolated locations	Iwapil	509.966	29	September 2006
	Reduction of environmental risks, posed by emerging contaminants, through advanced treatment of municipal and industrial wastes	Emco	1.199.987	36	June 2007
	Interaction between biology and membranes in membrane bioreactors	Biomem I	153.320	24	March 2008
	Integrated wastewater treatment process using mesh filter modules for direct activated sludge separation	Mesh	984.050	28	October 2008
	Water reclamation technologies for safe artificial groundwater recharge	Reclaim-water	3.560.440	39	December 2008
	Expanding membrane macroscale applications by exploring nanoscale material properties	NanoMemPro	6.380.000	54	February 2009
	MBR with an EU perspective for advanced municipal wastewater treatment strategies for the 21st century	Eurombra	2.998.969	44	May 2009
	Accelerate Membrane Development for Urban Sewage Purification	Amedeus	2.959.829	44	May 2009
	Reduction, modification and valorisation of sludge	Removals	2.936.470	36	June 2009
	New Energy Efficient approach to the operation of Membrane Bioreactors for Decentralised Wastewater Treatment	Puratreat	899.986	42	June 2009
	The Biomembrane: the development of a multifaceted skills base for the new millennium	Biomem II	2.548.323	48	August 2009
	Process optimisation and fouling control in membrane bioreactors for wastewater and drinking water treatment	MBR-Train	2.047.784	48	December 2009

(continued)

Table 4.1 Research projects membrane related funded within the 6th and the 7th FP (*Continued*)

FP	Project	Acronym	Funding (€)	Duration (months)	Completion
	Membrane-based desalination: an integrated approach	Medina	3.295.848	39	January 2010
	Innovative and integrated technologies for the treatment of industrial wastewater	Innowatech	2.750.000	41	March 2010
	Efficient membrane filtration units with integrated high frequency back pulsing device.	Hi-Fre	660.000	24	May 2011

Some of the EC funded projects within the 6th FP are briefly described below:

BIOMEM – Interaction Between Biology and Membranes in Membrane Bioreactors

This mobility project was held during 24 months and lasted until March 2008. In this fellowship the relation between biology and membrane fouling were studied from three perspectives: the determination of slime materials produced by the sludge, the microbial community composition of the sludge with special emphasis on the presence of slime-producing organisms and the development of a fouling measurement method which allows studying the process and mechanisms of membrane fouling. The latter was developed by the host institute for non-biological membrane filtration applications and was upgraded for MBRs.

The proposal went beyond the state of the art existing in that moment, by analyzing the relationship between biochemical sludge characteristics and membrane fouling for a wide range of operational conditions, wastewaters and MBR systems. It was highly innovative since no reports (until the proposal time) are known that try to find a relation between the composition of the microbial community in a MBR, the presence of specific organisms that may contribute to slime production, and membrane fouling.

Finally, the combined approach of biochemical analysis, microbial monitoring and application of a new and versatile fouling measurement method with the purpose of studying fouling phenomena is completely new for this application. The derived information and relationships had allowed adjusting MBR operation in such a way that membrane fouling is minimised. This has resulted in a lower frequency of membrane cleaning and an increased membrane lifespan. The reduced operational cost has enhanced MBR application for wastewater treatment and water reuse. The fellow has the background and the host institute has the infrastructure and the supporting team to make this multidisciplinary proposal a success.

MESH – Integrated Wastewater Treatment Process using Mesh Filter Modules for Direct Activated Sludge Separation

MESH was a research for SMEs project carried out during 44 months and ended in October 2008. The innovation of this project was to overcome the membrane drawbacks (high investments for membrane modules and high costs for membrane replacement as well as high operation costs due to increased energy demand) by substituting membranes by a cheaper filtration technique, offering higher flux rates at lower filter pressures.

Contact person

Dr. Werner Fuchs
Universität für Bodenkultur Wien – Department IFA-TULLN
Institute for Environmental Biotechnology
Austria
www.meshfilter.eu

NANOMEMPRO – Expanding Membrane Macroscale Applications by Exploring Nanoscale Material Properties

The NanoMemPro project was a joint initiative of 13 European partners, all leaders in the field of membrane technologies, coming from 13 member states, which aimed at:

- Setting up a 'European Membrane Technology Platform': shared research facilities.
- Promoting a common language: developing common procedures and protocols.
- Training researchers: basic training and continuous education. (European PhD and Master's Degree on Membrane Engineering are in preparation).

- Fostering European professional mobility.
- Creating a legal entity (European 'Membrane House' – AISBL) to host all the activities developed.
- Developing strong links with European industry.

After four and a half years, the NanoMemPro project came to an end in March 2009. The NanoMemPro programme's ambitious objectives were in large part reached, thanks to the efforts of all the various academic and industrial researchers taking part in the project.

Contact person

Prof. Gilbert M. RIOS
Institut Européen des Membranes (IEM-CNRS)
France
http://www.nanomempro.euromemhouse.com/

EUROMBRA – Membrane Bioreactor Technology (MBR) with an EU Perspective for Advanced Municipal Wastewater Treatment Strategies for the 21st Century

EUROMBRA is a research project developed during 44 months until May 2009. Its objective was to develop a cost-effective, sustainable solution for new, efficient and advanced municipal wastewater treatment based on MBR technology. This has been achieved through a multi-faceted, concerted and cohesive research programme explicitly linking key limiting phenomena (fouling, clogging) observed and quantified on the micro-, meso- and macro-scale. Key to the success of the programme is the harnessing specialist knowledge, conducting of dedicated yet interlinked experiments and incorporating key aspects of both, system design and operational facets, the latter encompassing hydrodynamics and mass transfer, foulant speciation and dynamic impacts.

Contact person

Orges Teknisk – Naturvitenskapelige Universitet
Faculty of Engineering Science and Technology, Department of Hydraulic and Environmental Engineering.
Norway

AMEDEUS – Accelerate Membrane Development for Urban Sewage Purification

AMEDEUS was a research project aiming at tackling two issues; (1) accelerating the development of competitive European MBR filtration technologies and (2) increasing acceptance of the MBR process through decreased capital and operation costs. It was developed during 44 months having its completion on May 2009.

The project targeted the two markets for MBR technology in Europe: the construction of small plants (semi-central, 50 to 2,000pe, standardized and autonomous), and the medium-size plants (central, up to 100.000pe) for plant upgrade. Technological development of new MBR systems were fostered by a consortium composed of 12 partners, of which five SMEs proposing novel concepts of low-cost and high-performance filtration systems. Two end-users, three non-profit institutions and two universities, all of them well versed in R&D in the MBR field, investigated solutions to reduce operation costs such as fouling control, membrane cleaning optimisation, aeration decrease, or optimise capital costs through improved implementation of membrane bioreactor process.

Furthermore, an analysis of the potential for standardisation was performed, and a technology transfer towards Southern and Eastern Europe was organised in order to facilitate the penetration of these new markets. AMEDEUS has achieved concrete and realistic technological breakthroughs for the MBR technology, and has improved the process engineering and operation practices. It has improved the competitiveness of the MBR European market and has rendered common this high-tech process for municipal wastewater treatment.

Contact person

Mr. Boris Lesjean
Kompententzzentrum Wasser Berlin Gemeinnutzige GmbH
Germany

PURATREAT – New Energy Efficient Approach to the Operation of Membrane Bioreactors for Decentralised Wastewater Treatment

PURATREAT is a specific targeted research project focussing on International Cooperation. It had a duration of 42 months and concluded on June 2009.

The objective of this project was to study a new approach to the operation of membrane bioreactors. This study included a comparison of the three leading hollow membrane technologies. The operating procedure studied expected to yield very low energy consumption and reduced maintenance costs. These characteristics would make the bioreactors working in these conditions suitable to be operated in peri-urban areas of the Mediterranean basin, where expenditure in public services is a critical factor.

Contact person

Dr. Gerhard Schories
ttz-Bremerhaven (Verein zur Förderung des Technologietransfers an der Hochschule Bremerhaven e.V.)
Water, Energy and Landscape Management
Germany
www.puratreat.com

BIOMEM II – The Biomembrane: the Development of a Multifaceted Skills Base for the New Millennium

BIOMEM is the second of two mobility projects aiming to enhance the knowledge base and mobility of researchers in biomembranes so as to generate effective, motivated and ambitious scientists with the skills to significantly improve out understanding of biological systems and through this, facilitate the development of innovative technologies and novel therapies. This first project lasted for 48 months and had its completion on August 2008.

The consortium which constitutes this training site took a holistic view of the biological membrane, by studying the relationships and fundamental features of proteins and their lipid environment in the context, where appropriate, of high quality projects. The various experts are grouped into co-ordinated clusters each of which was charged with delivering a training programme, integrated at Executive level. The scientific areas to be covered ranged from biochemistry to biophysics and encompassed a variety of horizontal skills such as Bioinformatics. The core components of the training site were designed to be complementary and addictive, thereby contributing to an accurate and meaningful description of the intact biological membrane. Finally, careful attention has been paid to the development of personal skills and individual monitoring, the overall object being the generation of multi-skilled flexible researchers equipped to meet the challenges that beckon.

Contact person

Dr. Heleen De Wever
Vlaamse Instelling Voor Technologisch Onderzoek
Environmental and Process Technology
Belgium

MBR-TRAIN: Process Optimisation and Fouling Control in Membrane Bioreactors for Wastewater and Drinking Water Treatment

MBR-TRAIN was a mobility action providing an early stage research training on process optimisation and fouling control in membrane bioreactors for water treatment. It was undertaken during 48 months and has been recently completed in December 2009.

As membrane fouling has been identified as a major barrier to sustainable MBR application, MBR-TRAIN undertook dedicated efforts to characterise and investigate both biological and physico-technical aspects of this phenomenon and developed strategies to control it. The findings have been used to advance the modelling of MBR processes in order to improve the prediction of plant performance and economic operation. The scope of the offered research project ranges from laboratory scale experimental set-ups to pilot- and full-scale operational plants.

The consortium of MBR-TRAIN comprises 10 partners from the water-industry, research institutes and universities across Europe representing a cross-section of relevant disciplines, sectors and regions. Due to its consortium composition, the MBR-TRAIN project provided an ideal framework for young researchers to prepare for future assignments in intersectorial tasks pursuing a research career in both academic institutions and industrial enterprises. MBR-TRAIN offered the young researchers a well organised and structured training programme in a technology field of growing relevance. They were benefited from a tutored PhD thesis work with additional scientific training offered in short-courses or workshops. The exchange of personnel between the participating institutions has trained adapting to work in different research structures. In order to bring benefit to the researchers careers beyond a merely scientific formation, MBR-TRAIN has provided training on complementary skills such as presentation techniques, communication skills, intellectual property rights or research project management.

Contact person

Prof. Thomas Melin
Rheinisch Westfaelische Technische Hochschule Aachen.
Institut Fuer Verfahrenstechnik – Chemical Engineering Department
Germany
www.mbr-train.org

MEDINA – Membrane-based Desalination: an Integrated Approach

MEDINA was a research project developed during 39 months that was successfully completed on January 2010. The aim of this project was to solve the issues affecting RO technology; such as, improvement of water quality, enhancement of the recovery factor, reduction of the unit water cost, minimizing the brine disposal impact, an innovative approach based on the integration of different membrane operations in pre-treatment and post-treatment stages is proposed. The expected outcomes and contributions of the research were:

- The development of advanced analytical methods for feed water characterization, appropriate fouling indicators and prediction tools, procedures and protocols at full-scale desalination facilities.
- Identification of optimal seawater pre-treatment strategies by designing advanced hybrid membrane processes (submerged hollow fibre filtration/reaction, adsorption/ion exchange/ozonation) and comparison with conventional methods.
- The optimisation of RO membrane module configuration, cleaning strategies, reduction of scaling potential by NF.
- The development of strategies aiming to approach the concept of Zero Liquid Discharge (increasing the water recovery factor up to 95% by using Membrane Distillation – MD; bringing concentrates to solids by Membrane Crystallization or Wind Intensified Enhanced Evaporation) and to reduce the brine disposal environmental impact and cost.
- To increase the sustainability of desalination process by reducing energy consumption (evaluation of MD, demonstration of a new energy recovery device for SWRO installations) and use of renewable energy (wind and solar).

The research team embodied science and engineering from both the practitioner and academic perspectives. Potential end-users and participating utilities were involved in research activities and applications. Link ages with ongoing research activities and demonstration studies at full-scale desalination plants were conducted to ensure the applicability and transfer of the findings of the proposed research project.

Contact person

Prof. Enrico DRIOLI
Department of Chemical Engineering and Materials
Italia
http://medina.unical.it/

HI-FRE: Efficient Membrane Filtration Units with Integrated High Frequency Back Pulsing Device

The objective of this project is to overcome technological limitations of membranes (fouling phenomena and pore blocking) by developing, a high frequency back-pulsing device, integrating it into membrane filtration systems and transferring the knowledge gained to new, more competitive products and services offered by the SMEs involved. This is a research for SMEs project that will be running during 24 months and will have its end on May 2011.

The key feature of the innovative back-pulse concept is a valve-less construction providing short response times, defined pulse shape and efficient membrane cleaning at minimal back-pulse flow. Selected membranes (polymeric and ceramic) and modules (capillary and flat sheet) will be tested. The integration of pulsing device and module is essential for successful scale up. It has to take into account inertia, viscosity and elasticity effects and gets prime attention in a dynamic modelling approach. The applications to be investigated range from the treatment of liquid residues of biomass-based power generation to treatment and reuse of process fluids and wastewater, including membrane bioreactor applications.

The project addresses all critical points along the value chain from membrane supply to end-use. Its outputs include insight into the hydrodynamics of high frequency back-pulsing, novel back-pulsing devices, adapted membrane/module configurations and new applications for a new technique. The consortium includes 2 RTD partners focusing on technology, modelling and testing, 4 SMEs manufacturing back-pulsing devices, ceramic membranes, polymeric membranes and filtration systems, 1 SME as system integrator, 2 SME end-users and one large end-user for demonstration.

Contact person

Ms. Ulrich Brüss
A3 Water Solutions GmbH
Germany

4.2 MBR CLUSTERS

4.2.1 MBR-Network

The European Commission has decided to promote the development of the MBR technology while funding four projects entirely dedicated to research, development, capacity building and technological transfer in regards to this promising wastewater treatment process. This four projects previously described, namely AMEDEUS, EUROMBRA, MBR-TRAIN and PURATREAT were supported by three different financial instruments set up by the European Commission within the sixth Framework Program, and were implemented in parallel from October 2005 up to December 2009.

Around 50 European and international companies and institutions were actively involved in these four projects and have joined their efforts and coordinated their actions within the cluster 'MBR-Network'. The four projects amounted to a total budget of ca. 15 Mio€, for which ca. 9 Mio€ were financed by the European Commission. They represent the largest coordinated research initiative worldwide dedicated to MBR technology since the first developments of this treatment process in the early 90's.

Important technological breakthroughs, process improvement, knowledge and capacity transfer and building are expected, which will lead to better acceptance, competitiveness and broader implementation of the technology in both municipal and industrial fields.

4.2.2 IWA Membrane Technology Group

The International Water Association (IWA) is the global network of 10,000 water professionals spanning the continuum between research and practice and covering all facets of the water cycle.

The objectives of the group are to advance knowledge and experience in the area of development and implementation of membrane technology for drinking and industrial water production as for low quality and wastewater treatment. It is associated with the development of membranes and adjustment to the low quality to be treated and subject to the required qualities.

The group promotes membrane technology for drinking and industrial water production as for low quality and wastewater treatment in both domestic, industrial and agricultural applications; The group develops and encourages international cooperation towards optimal membrane performance and performance evaluation; support operational research studies and management modelling of membrane systems; identifies research needs; and develops international technology transfer processes related to the use of membranes for improving the processes of water and wastewater treatment in relation with the regulations.

4.2.3 European Membrane House

The European Membrane House is a non-profit making international association set up in 2008 by universities and research institutes in 10 European countries working together within the framework of NanoMemPro, already commented above. The European Membrane Society is also an EMH founding member.

Strategic business and research agenda (SBRA, made in 2007 within the framework of NanoMemPro) set out research and technological development priorities for the mid to long term as reflected by a wide range of stakeholders. Their membrane SBRA clearly brings to light the sectors in which membranes are one of the best techniques available, what technological breakthrough is awaited. It also points out what needs to be improved for membranes to be more user-friendly and is a tool to bring together industrialists and academic researchers in valuable collaborative R&D projects.

4.2.4 UNESCO Centre for Membrane Science and Technology

The Centre has five academic and over 20 researchers dedicated to membrane science and technology. The Centre has a strong profile and recognition factor internationally as one of the largest membrane groups and as a world leader in a wide range of research areas.

Much of the Centre's reputation rests on their approach to wider generic problems from a fundamental engineering science approach incorporating skills from physical chemistry to high level computing, rather than solely from an application focus. Historically, it has been a University Centre for over 20 years and was previously supported as Commonwealth Special Research Centre for 9 years.

Recently, the Centre has leveraged its collective infrastructure and expertise to form collaborative links with other Centres and schools and participate in major research initiatives nationally and internationally. These include a major

CSIRO cluster in water, 3 DEST International Linkage Programmes, and three Cooperative Research Centres. This has brought significant research income and profile to UNSW in addition to their continuing success with traditional sources of funding from the Australian Research Council and industry. Due to the increasing role of membranes in the global issues such water, energy, and climate change, the Centre is in a position to take a research leadership role in many of these key technology challenges.

Website: http://www.membrane.unsw.edu.au/index.asp

4.3 REFERENCES

Cordis. http://cordis.europa.eu/home_en.html (accessed 4 March 2010).
European Membrane House. www.euromemhouse.com/ (accessed 4 March 2010).
IWA Membrane Technology Group. www.iwahq.org/Home/Networks/Specialist_groups/List_of_groups/Membrane_Technology/.
MBR Network. www.mbr-network.eu/ (accessed 4 March 2010).
UNESCO Centre for Membrane Science and Technology. www.membrane.unsw.edu.au/index.asp (accessed 4 March 2010).

Chapter 5

Conclusions and recommendations

Gerhard Schories, Rashed Al-Sa`ed and Alessandra Criscuoli

Water scarcity is a serious problem in the MENA countries. Water of certain quality is needed for agriculture (food production), to operate industrial facilities, tourism, household purposes, and last but not least simply for drinking purposes. It is obvious, that scarce conventional water resources cannot meet a steadily increasing demand any longer. One important option to solve this problem is the production of potable water from seawater by means of desalination. An alternative which is becoming more and more important is the treatment and reuse of wastewater. Both ways are possible and in both concepts membranes (in different technical configurations) play the key role.

Specific research, but also individual commitment and engagement are necessary for the successful implementation of membrane technology for water treatment in the MENA countries. It is necessary to develop high efficiency systems and tailor made solutions. The first step for ProMembrane was the determination of the actual status of R&D activities in the target region. Several research initiatives have been identified, both, on a local but also on an international level.

The first initiatives have been carried out already in the frame of ProMembrane. Two international fairs have been organised in Syria and Egypt, respectively, in order to inform local stakeholders about membrane technology and its technical potentials. In scientific lectures in parallel to the technical exhibitions a discussion between local researchers and industrials with European experts has been started. Additionally, an international conference in Tunisia with more than 800 international participants has proven the high relevance and importance of membrane technology for water treatment for reuse in the MENA region.

For sustainability it is very important to encourage young researchers and engineers to work with and on membrane technology. As a starting signal, ProMembrane has set an award for young scientists to honour actual research work and to motivate the young to carry out membrane based R&D in the future. Quite an impressing number of applications for the 'ProMembrane Young Scientist Award' had been submitted. Although the scientific quality of the applications was varying in a wide range, all of them deserve to being supported, as they all contribute significantly to an improvement of the water treatment and supply situation in the target region, not only the three applications, that finally won the awards.

However, much more applied research is required to cope with actual and upcoming problems to provide water in sufficient amount and good quality to everyone in the MENA region, to farmers, industry, tourism and private households. Such research work should be carried out in local and international cooperation. The EU has financially supported such R&D cooperation by means of the past Research Framework Programs and additional instruments such as Europe Aid. An overview of the most relevant R&D projects has been listed in this publication including

 Membrane Technology in Water Treatment in the Mediterranean Region (ProMembrane). By Antonia Lorenzo and Anibal Vega. ISBN: 9781843393702. Published by IWA Publishing, London, UK.

contact details of the consortia in order to enable the reader of this publication to get in touch with these projects. The information is also available on the project website (www.promembrane.info). Several are ongoing; more projects will come up in the current 7th Research Framework Program of the EU.

Besides the activities carried out so far, much more initiatives are required. The following aspects of membrane technology should be in the focus:

Strategic objectives

- Centralised vs decentralised concepts

Technical and environmental objectives

- Purification efficiency
- Mode and reliability of operation
- Energy efficiency
- Application of renewable energy sources (e.g. wind, solar energy) for energy supply

Economic objectives

- Cost effectiveness
- Local availability of components and services

Social objectives

- Further awareness rising
- Training and capacity building

Technology transfer, training and capacity building at the end of R&D projects secure the successful and sustainable implementation of the projects' outcomes.

Another important approach to intensify and extend R&D on membrane technology for water treatment is the establishment of local centres of excellence for steering and coordinating research activities on local bases. Successful examples for such centres can be found e.g. on the Arab peninsula. The Price Khalid Bin Sultan Chair for Water Reuse Research at King Saud University, Riyadh, Saudi Arabia (http://pkc-wreuse.ksu.edu.sa/) and the Middle East Desalination Research Centre in Muscat, Oman (www.medrc.org) have to be mentioned in this context. The latter encourages cooperation between local and international researchers and provides also financial support, whereas the Price Khalid Bin Sultan Chair for Water Reuse Research initiates international cooperation through workshops etc.

To summarize, several important joint activities have been started. PROMEMBRANE as one of those joint EU-MENA activities funded by the EC provides a platform for collecting and sharing information as well as for finding cooperation partners, locally and all over the world. This platform shall foster and facilitate future R&D activities in Africa, Middle East and beyond. A lot more work needs to be done to secure supply of water of good quality in a water scarce region with rapidly growing population. By successfully completing PROMEMBRANE an important milestone has been reached to promote, encourage and support future research in the sector.

Annex I

Database of documents

ProMembrane consortium

The following tables are part of the database of documents developed within the ProMembrane project. In the tables 159 references of papers are included. These papers have been published in Algeria, Cyprus, Egypt, France, Germany, Greece, Italy, Jordan, Libya, Palestine, Spain, Syria, Tunisia and Turkey.

Information on the authors, year and country of publication, journal name, funding agent, technology used and use (water treatment (WT) and wastewater treatment (WWT)), are given. In addition, an abstract and some recommendations are also included.

Authors	Year	Title	Journal name, volume (issue)	Abstract	Country	Funding Agent	Technology used	USE (WT/WWT)	Recommendations
Bouchekima, B., Gros, B., Ouahes, R., Diboun, M.	2001	Brackish water desalination with heat recovery	Desalination 138 (2001) 147–155	The aim of our research is to develop a highly efficient desalination system of simple construction, applicable in units of less than $1m^3/d$ of flesh water production, suitable also for areas with poor infrastructure.	Algeria	NA	Solar desalination	WT/BW	The experimental results obtained show the significant superiority of this type of distiller over the conventional basin-type solar still.
Benzaoui, A., Bouabdallah, A.	2004	Desalination and biological wastewater treatment process	Desalination 165 (2004) 105–110	This paper relates to an experimental investigation about the treatment processes of large quantities of used water from agricultural uses.	Algeria	NA	Biological treatment	WWT/ domestic	The realization of such a project is ecologically very interesting with regard to the conservation of natural resources for future generations and for the well-being of the present population.
Badania, Z., Ait-Amara, H., Si-Salahb, A., Brikc, M., Fuchsc, W.	2005	Treatment of textile waste water by membrane bioreactor and reuse	Desalination 185 (2005) 411–417	The objective of this study is to determine the favourable operating conditions to the functioning of the membrane bioreactor with external buckle for the variable operating conditions to estimate the capacity of degradation, the quality of the treated water, the criteria of conception concerning the size of the reactor and the energy consumption.	Algeria	NA	MBR	WWT domestic	Long-terms performance of a tangential flow MBR for textile wastewater treatment proved the average removal COD as 97%. A hydraulic retention time in the bioreactor of 1,2 day was sufficient to obtain degradation of COD when operating at a sludge concentration of 15g/L. The average removal colour was 70%. We recommend a post-processing in nanofiltration.
Belkacem, M., Bekhti, S., Bensadok, K.	2007	Groundwater treatment by reverse osmosis	Desalination 206 (2007) 100–106	An experimental studies was undertaken over 12 months in a two reverse osmosis	Algeria	NA	RO	WT/ Ground water	The performances of reverse osmosis process treatment for production of ultrapurewater were examined in terms of ion

(*continued*)

Authors	Year	Title	Journal name, volume (issue)	Abstract	Country	Funding Agent	Technology used	USE (WT/WWT)	Recommendations
				module in the beverage industry, the groundwater was treated by reverse osmosis to eliminate the water solutes.					rejection and removal of microorganisms.
Truby, R.	2001	Desalination processes enhanced by multiple membrane systems	EDS Newsletter Issue 12, May 2001	Desalination using RO has been an established process for many brackish and seawater applications for over three decades and is found in many market sectors.	Cyprus	NA	UF MF RO	WT/Sea water	The reliability of the RO is greatly increased as a result of consistent 'clean' feed water provided by the UF or MF membrane. Cleanings and downtime are reduced. Capital costs of the RO can be reduced significantly when UF or MF is used as pre-treatment.
Koutsakos, E., Savvides, K., Savva, K.	2005	Larnaca desalination plant operation – a client and contractor perspective	Desalination 184 (2005) 157–164	The plant operation from two perspectives: (a) the monitoring activities and obligations set by the client in order to make sure that the plant satisfies the contractual criteria; (b) the plant performance achieved by the contractor in relation to the contractual obligations from commissioning up to date (2001–2004).	Cyprus	NA	RO	WT/SW	Avoid if possible commissioning desalination plants in warm – summer months. Produce clear and well defined Contract. A good contractor, client laboratory, co operation at all levels of client and contractor is essential – management, communications with all stakeholders of the plant.
Atikol, U., Aybar, H.S.	2005	Estimation of water production cost in the feasibility analysis of RO systems	Desalination 184 (2005) 253–258	In this study a practical method is developed where the recent experiences are modified according to the local conditions for making reasonable estimations of the unit production cost of fresh water for new RO systems to be constructed.	Cyprus	NA	RO	WT/SW	A method of using a comparative analysis of cost estimation was developed in the feasibility of RO plants for areas where the RO plants were never installed before.

(*continued*)

Authors	Year	Title	Journal name, volume (issue)	Abstract	Country	Funding Agent	Technology used	USE (WT/WWT)	Recommendations
Basta, A.H., El-Saied, H., Elberry, M.	2003	Cellulose membranes for reverse osmosis Part II. Improving RO membranes prepared from non-woody cellulose	Desalination 159 (2003) 183–196	The use of a heterogeneous acetylation process is preferable in preparing CA membranes with relatively high degree of desalination	Egypt	NA	RO	WT/Brackish water	The produced water from double passing the brackish water through the regenerated cellulose 9.1% PP, followed by passing it through regenerated celluloses from methylol cellulose, A-membranes and r COOMA-membranes could be used for drinking and domestic purposes.
AL-Manharawy, S., Hafez, A.	2003	A new chemical classification system of natural waters for desalination and other industrial uses	Desalination 156 (2003) 163–180	The aim of this work was to develop a water chemical classification system that can provide a useful tool to differentiate, characterize and classify the different nature surface and ground water over wide range of salinity.	Egypt	NA	RO	WT/Sea water	The chemical identification and description of different water kinds become easier and more accurate; this provides a valuable techno-economical tool for proper selection and development of specific water treatment methods, chemicals, equipment and material that are suitable for desalination technology and other industrial uses.
EL-Zanati, E., Eissa, S.	2004	Development of a locally designed and manufactured small-scale reverse osmosis desalination system	Desalination 165 (2004) 133–139	This article presents the design principles and features, as well as the failure of the configuration and its causes; also, the development of the first prototype is discussed.	Egypt	NA	RO	WT/Sea water	The water production costs were decreased by simplification of the system.
Abdel-Shafy, H.I., Hegemann, W., Guindi, K.A., Badawy, M.I., Tawfik, N.S., Teschner, K.	2005	Membrane Bioreactor for the Treatment of Wastewater in an Egyptian Plant	CEJOEM 2005, Vol. 11. No. 3.: 217–223	The membrane bioreactors (MBRs) are strongly developing new concepts for wastewater treatment warranting the fulfilment of high hygienic standards for the effluent.	Egypt	NA	MBR	WWT/Municipal	The results obtained showed significant improvement in the quality of total suspended solids, biochemical oxygen demand, nitrogen compounds, phosphates, electric conductivity, dissolved oxygen, and heavy metals.

(continued)

Authors	Year	Title	Journal name, volume (issue)	Abstract	Country	Funding Agent	Technology used	USE (WT/WWT)	Recommendations
El-Zanati, E., El-Khatib, K.M.	2007	Integrated membrane – based desalination system	Desalination 205 (2007) 15–25	A comparative study of an integrated hybrid membrane – based system with an earlier locally designed RO unit, such system comprises of Nanofiltration (NF), Reverse osmosis (RO) and Membrane Distillation (MD) subsystems.	Egypt	NA	RO, NF & MD	WT/SW	The integration of many membrane operations improves the performance of seawater desalination unit. The data demonstrates that the water cost is competitive compared to those of potable water produced in SWRO system.
Mabrouk, A.A., Nafey, A.S., Fath, H.E.S.	2007	Thermoeconomic analysis of some existing desalination processes	Desalination 205 (2007) 354–373	This paper presents a thermoeconomic analysis of the widely used and existing desalination processes. Thermoeconomic approach is used to distribute the cost of the whole process on the internal streams based on exergy not energy.	Egypt	NA	MSF, MEE, MEE-TVC, MEE-MVC & RO	WT/SW	Comparison among the considered processes shows that the unit product cost of RO process is the lowest. The MEE-MVC is in competition with RO process if the reliability and long life of the heating surface are taken into consideration.
Khordagui, H.K.	1995	Fate and control of nerve chemical warfare agents in the desalination industry of the arabian-persian gulf	Environment International, Vol. 21, No. 4, pp. 363–379, 1995	This paper assesses the impact of chemical warfare agents (CWAs) on drinking water. The study is focussed on the Arabian-Persian Gulf (APG) and the desalination process.	Egypt	NA	RO	WT/SW	Several major gaps existed in the knowledge needed to accurately determine the fate of nerve CWAs and their impact on desalinated drinking water quality in the A.P.G.
Ahmed, S.A., Tewfik, S.R., Talaat, H.A.	2002	Development and verification of a decision support system for the selection of optimum water reuse schemes	Desalination 152 (2002) 339–352	This paper addresses the development and verification of an expert system for the intelligent selection of the optimal scheme for the treatment of agricultural drainage water.	Egypt	NA	RO	WWT/domestic	It is recommended that this approach would be complemented with process design routines to provide preliminary process designs and cost estimates for a specific case under consideration. • extended to optimize integrated management and reuse schemes for a geographical area.

(*continued*)

Authors	Year	Title	Journal name, volume (issue)	Abstract	Country	Funding Agent	Technology used	USE (WT/WWT)	Recommendations
Shaalan, H.F.	2002	Development of fouling control strategies pertinent to nanofiltration membranes	Desalination 153 (2002) 125–131	This paper presents a performance analysis of some commercial NF membranes via reported flux decline data.	Egypt	NA	NF	WT/Surface water	The performance of NF systems is governed by membrane characteristics, water compositions and operating conditions. Published data have been managed and analyzed to come up with empirical correlations describing time dependence of normalized flux.
Hafez, A., El-Manharawy, S.	2002	Economics of seawater RO desalination in the Red Sea region, Egypt. Part 1. A case study	Desalination 153 (2002) 335–347	The present work deals with the economic evaluation and limitation of the small size SWRO plants, which are widely used in the new tourism development areas at South Sinai and the Red Sea coast.	Egypt	NA	RO	WT/SW	It is recommended to study and investigate the brine disposal into the suitable deep geological formation. It is recommended to avoid the highly salty groundwater (TDS >50,000mg/l) by using the surface intake (TDS – 43,000mg/l) instead. It is recommended to rearrange the tourism development strategy and regional planning as well. It is recommended to reconsider the individual installation of the costly small SWRO plants.
El-Saied, H., Basta, A.H., Barsoum, B.N., Elberry, M.M.	2003	Cellulose membranes for reverse osmosis Part I. RO cellulose acetate membranes including a composite with polypropylene	Desalination 159 (2003) 171–181	RO membranes were prepared from chemically modified non-woody cellulosic materials (purified cotton linters and krafi bagasse pulp) for desalination of brackish water.	Egypt	NA	RO	WT/Brackish water	The membrane properties prepared from conventional acetylated celluloses have relatively higher performance than those obtained from non conventional acetylated celluloses when both are compared with the standard purchased cellulose acetate.
Elmaleh, S., Abdelmoumni, L.	1997	Cross-flow filtration of an anaerobic methanogenic suspension	Journal of Membrane Science 131 (1997) 261–274	The purpose of this work is then to study the filtration of a methanogenic suspension produced in	France	H.B.rral	MF, UF	WWT/Municipal	The next step should be the study of the filtration of a suspension produced in an anaerobic reactor where the carbon source is

(continued)

Authors	Year	Title	Journal name, volume (issue)	Abstract	Country	Funding Agent	Technology used	USE (WT/WWT)	Recommendations
				an anaerobic reactor where the sole organic substrate is acetic acid.					a mixture of volatile fatty acids.
Couffin, N., Cabassud, C., Turcaud, V.L.	1998	A new process to remove halogenated VOCs for drinking water production: vacuum membrane distillation	Desalination 117 (1998) 233–245	The present paper aims to evaluate if VMD could be used to remove HOVs at very low concentration from water from experimental results obtained with synthetic waters.	France	Seine-Normandie Water Agency for financial support.	Vacuum membrane distillation	WT/DW	Further studies will focus on the comparison of different commercial available membranes for this application and also on a complete study about the influence of operating conditions which will include the feed velocity.
Tardieu, E., Grasmick, A., Geaugey, V., Manem, J.	1998	Hydrodynamic control of bioparticle deposition in a MBR applied to wastewater treatment	Journal of Membrane Science 147 (1998) 1–12	This paper aims to examine whether there are any operating conditions capable of limiting the deposition of bioparticles.	France	NA	MBR	WWT/Activated sludge	Therefore it is vital to pursue research into the interactions between the biological conditions of the process and the filtration performances. Moreover, it will make it possible to determine the critical flux value as a function of hydrodynamic and biological conditions.
Pontie, M.	1999	Effect of aging on UF membranes by a streaming potential (SP) method	Journal of Membrane Science 154 (1999) 213–220	In this paper we have determined, by a streaming potential (SP) method, the isoelectric point (IEP) of a new cellulose membrane, the regenerated cellulose material.	France	Ecole Nationale SupeÂrieur de Chimie de Paris	UF	WT	We are currently investigating this subject and the SP measurements to a larger surface for different kinds of materials.
Tardieu, E., Grasmick, A., Geaugey, V., Manem, J.	1999	Influence of hydrodynamics on fouling velocity in a recirculated MBR for wastewater treatment	Journal of Membrane Science 156 (1999) 131–140	Membrane fouling within membrane bioreactor (MBR) applied to wastewater treatment was studied under adequate recirculation conditions, for which fouling was only caused by soluble and	France	NA	MBR	WWT/Municipal	Many other questions remain to be answered: Just how reversible is the fouling structure formed (by cancelling out the TMP or recirculating water)? What is the fouling structure's morphology? What kind of physico-chemical interactions is it

(*continued*)

Authors	Year	Title	Journal name, volume (issue)	Abstract	Country	Funding Agent	Technology used	USE (WT/WWT)	Recommendations
				colloidal fractions of the biological suspension.					structured by? What is its chemical composition? What is the biological origin of the involved compounds? How does the composition of the fouling structure vary in relation to the biological conditions? What are the overall working parameters that are likely to affect the rate of fouling? The modelling shown here constitutes a helpful tool for studying these points in more detail, in the future.
Bequet, S., Abenoza, Aptel, T.P., Espenan, J.M., Remigy, J.C., Ricard, A.	2000	New composite membrane for water softening	Proceedings of the Conference on Membranes in Drinking and Industrial Water Production, Volume 1, pages 495–501, 2000	In this paper, we report about another route to the formation of nanofiltration membrane by modification of ultrafiltration membranes with grafted polymer. This was accomplished by UV irradiation.	France	Région Midi Pyrénéees and by the Centre National de la Recherche Scientifique	NF	WT	This preliminary study will be continued to clarify multiple points that have been raised during this work. To achieve the objective it is necessary to determine ultrafiltration membrane characteristics grafting yield and the grafted layer structure (inside the membrane and on the surface).
Heran, M., Elmaleh, S.	2001	Prediction of cross-flow micro-filtration through an inorganic tubular membrane with high-frequency retro filtration	Chemical Engineering Science 56 (2001) 3075–3082	This paper intends to evaluate the accuracy of the published models by observation of transient signals with two different suspensions.	France	NA	MF	W&WWT/ (bentonite in tap water and biologically treated wastewater)	The introduction of a critical time for the deposit is not necessary. The cleaning efficiency is an important parameter, which can be experimentally determined. Model II predicts correctly the net flux and the optimal frequency. No model predicts accurately the reverse flux when the particles interact with the membrane. The response to one backpressure pulse

(*continued*)

Authors	Year	Title	Journal name, volume (issue)	Abstract	Country	Funding Agent	Technology used	USE (WT/WWT)	Recommendations
									allows predicting the performance of high-frequency retrofiltration.
Romero, J., Cam, S.L., Sanchez, J., Saavedra, A., Rios, G.M.	2001	Permeation of supercritical fluids through a MFI zeolite membrane	Chemical Engineering Science 56 (2001) 3139–3148	The objective of this study is the determination and description of transfer mechanisms of supercritical fluids through inorganic microporous membranes whose selective layer is constituted of MFI zeolite, a material with particular properties of porosity, crystalline structure and adsorption.	France	This publication is the result of a cooperation programme between the University of Montpellier and the University of Santiago de Chile, supported by a grant from the French Government to the chilean student Julio Romero.	MFI zeolite membrane	Supercritical fluids (SCF) and inorganic microporous materials.	Starting from literature data, model parameters have been identified. It appears that the energy of activation for diffusion is the same whatever the membrane and the values are of the same order of those obtained by other authors.
Brehant, A., Bonnelye, V., Perez, M.	2002	Comparison of MF/ UF pretreatment with conventional filtration prior to RO membranes for surface seawater desalination	Desalination 144 (2002) 353–360	The objective of the study was to assess the potential of ultrafiltration (UF) pre-treatment prior to reverse osmosis (RO) for desalting seawater with high-fouling tendency.	France	NA	MF, UF&RO	WT/SW	The performance of RO membranes downstream UF exceeded the usual operating conditions encountered in seawater desalination. The combined effect of higher recovery and higher flux rate promises to significantly reduce the RO plant costs. Longer runs are now required to confirm these preliminary operating data.
Vial, D., Doussau, G.	2002	The use of microfiltration membranes for seawater pre-treatment prior to reverse osmosis membranes	Desalination 153 (2002) 141–147	This paper describes pilot test results obtained on Mediterranean seawater using a 0.1m hollow fiber Microza® module.	France	NA	MF&RO	WT/SW	Long-term seawater pilot testing has shown the 0.1μm Microza hollow-fiber module's ability to provide clarified water for RO desalination systems.
Jorda, J., Marechal, P., Rigal, L., Pontalier, P.Y.	2002	Biopolymer purification by ultrafiltration	Desalination 148 (2002) 187–191	The aim of this work was to analyze the role of the membrane configuration and of	France	NA	UF	WWT/Industrial (Biopolymer)	This study shows ultrafiltration could be introduced in biopolymer purification process, but

(*continued*)

Authors	Year	Title	Journal name, volume (issue)	Abstract	Country	Funding Agent	Technology used	USE (WT/WWT)	Recommendations
				the MWCO on the filtration performances (flux and loss) for the concentration of arabinoxylan and rhamnogalacturonan solutions.					many points have to be improved. Membrane MWCO has to be optimized for arabinoxylans filtration to avoid fouling. Trials are realized with higher MWCO until 50kD. Influence of shear on rhamnogalacturonan structure has also to be studied with different devices.
Ognier, S., Wisniewski, C., Grasmick, A.	2002	Characterisation and modelling of fouling in membrane bioreactors	Desalination 146 (2002) 141 – 147	The objective of this work was (i) to ensure that stable filtration conditions could be obtained in an MBR under constant hydrodynamic, biological and physico-chemical conditions and (ii) to study the influence of weak variations of two physicochemical parameters (pH and temperature) on process stability.	France	NA	MBR	WWT/Municipal	In an MBR for denitrification, the great alkalinity of the suspension can cause the precipitation of calcium carbonate for pH values between 8 and 9. The role of precipitation can be pointed out as a cause of system instability, even if the system works in subcritical conditions. The fouling mechanism could be the deposition of Cacao, particles formed in the bulk suspension by bulk crystallisation. Further research would be necessary to confirm this hypothesis.
Chatila, F.	2003	Nanofilters Convert River Water into Drinking Water: French City Finds a New Source for its Citizens	Water Conditioning & Purification, July 2003	In northern Paris, the Syndicate des Eaux d'Ile de France (SEDIF) supplies 39 communities (800,000 people) from a plant that employs a very unique method of treating water from a river source. Located in Méry-sur-Oise.	France	Syndicate des Eaux d'Ile de France (SEDIF)	NF & RO	WT/RW	People receive drinking water without any taste of chlorine and soft water that no longer causes any scaling problems in home plumbing.

(continued)

Authors	Year	Title	Journal name, volume (issue)	Abstract	Country	Funding Agent	Technology used	USE (WT/WWT)	Recommendations
Nisan, S., Commerc, B., Dardour, S.	2005	A new method for the treatment of the reverse osmosis process, with preheating of the feed water	Desalination 182 (2005) 483–495	This paper summarises our investigations on the conceptual studies of the Reverse Osmosis (RO) process with preheating of the feed-water, which is expected to lead to lower specific power consumption and higher water production, thus further reducing the cost of desalination by this advanced RO process.	France	Filmtec/Dow Chemical Company	RO	WT	The basic aim of investigations presented in this paper was to understand the behaviour of the RO system, with the preheating of the feed-water. Economic studies have shown that ROPH could be a very interesting method for reducing desalination cost.
Besslere, Y., Bacchin, P., Jefferson, B.	2005	Dead-end filtration of natural organic matter: experimental evidence of critical conditions	Desalination 175 (2005) 29–36	The aim was to investigate the idea that it may be possible to avoid irreversible fouling formation in dead-end filtration, even if the suspension studied contains organic material.	France	Cranfield School of Water Science. University Paul Sabatier and Aquasource	Dead-end ultrafiltration cell	WT/NOM	The results indicate that for a given pressure there exist an associated critical filtered volume, below which fouling is reversible by rinsing whereas above this critical filtered volume a significant part of fouling is irreversible.
Massé, A., Spérandio, M., Cabassud, C.	2006	Comparison of sludge characteristics and performance of a submerged membrane bioreactor and an activated sludge process at high solids retention time	Water Research 40 (2006) 2405–2415	This work aims to compare biomass structure and performance of a submerged membrane bioreactor (SMBR) and an activated sludge process (ASP) treating the same domestic wastewater	France	Aldo Moro, Evrard Mengelle, Bernard Reboul and Michel Mauret	MBR	WWT/domestic	Membrane separation and also modification of F/M ratio and SRT induce modification of sludge characteristics which are of great importance for predicting physico-chemical properties of the sludge, and these findings will need to be correlated to fouling tendency in further work.
Grelier, P., Rosenberger, S., Tazi-Pain, A.	2006	Influence of sludge retention time on membrane bioreactor hydraulic performance	Desalination 192 (2006) 10–17	The objective of this study was to investigate the influence of the sludge retention time on the operation performance of the membrane bioreactors.	France	NA	MBR	WWT/Municipal	Since the filtration performance is one of the main economical factors in MBR operation, the optimal sludge retention time should be balanced between 15 and 40d, according to the results of this study.

(*continued*)

Authors	Year	Title	Journal name, volume (issue)	Abstract	Country	Funding Agent	Technology used	USE (WT/WWT)	Recommendations
Rosenberger, S., Laabs, C., Lesjean, B., Gnirss, R., Amye, G., Jekel, M., Schrotter, J.-C.	2006	Impact of colloidal and soluble organic material on membrane performance in membrane bioreactors for municipal wastewater treatment	Water Research 40 (2006) 710–720	Two parallel membrane bioreactors ($2m^3$ each) were operated over a period of 2 years. Both pilots were optimised for nitrification, denitrification, and enhanced biological phosphorous elimination, treating identical municipal wastewater under comparable operating conditions.	France	EC	MBR	WWT/Municipal	The presented paper proves the correlation between non-settleable substances and membrane filtration at a reasonable scale, but work on the interactions on the membrane surface is still needed for a comprehensive understanding of the processes and a consecutive process optimisation.
Jacob, C.	2007	Seawater desalination: Boron removal by ion exchange technology	Desalination 205 (2007) 47–52	Ion exchange is the most appropriate technology to remove boron selectively from desalinated seawater.	France	NA	Ion exchange & RO	WT/SW	Ion exchange technology demonstrates significant benefits for removing boron in a desalination application. Thanks to its high selectivity, ion exchange resin performance is not impacted by operating conditions such as temperature, pH, or salinity.
Euvrard, M., Hadi, L., Foissy, A.	2007	Influence of PPCA (phosphino-polycarboxylic acid) and DETPMP (diethylene-triamine-pentamethylene-phosphonic acid) on silica fouling	Desalination 205 (2007) 114–123	This paper is focused on silica formation, deposition and its inhibition in desalination applications by following the kinetic of silica formation and morphometric characteristics of deposits.	France	NA	Desalination	WT/Industrial	This study carried on the inhibition performance of chemical inhibitors – phosphino-polycarboxylic acid (PPCA) and diethylene-triamine-pentamethylene-phosphonic acid (DETPMP) on silica fouling. Experimental results showed high effects on the inhibition efficiency of these molecules as soon as their concentrations reached 50ppm: The deposition of silica was highly reduced.

(*continued*)

Authors	Year	Title	Journal name, volume (issue)	Abstract	Country	Funding Agent	Technology used	USE (WT/WWT)	Recommendations
Bonnélye, V., Sanz, M.A., Francisci, L., Beltran, F., Cremer, G., Colcuera, R., Laraudogoitia, J.	2007	Curacao, Netherlands Antilles: A successful example of boron removal on a seawater desalination plant	Desalination 205 (2007) 200–205	The paper presents the first results of the plant optimization and the performance tests in terms of water quality, RO membrane performances based on the initial project and the site figure during the first months of operation.	France	Aqualectra	RO	WT/SW	The treatment plant first results are in accordance with the expected values of water quality. Some preliminary adjustments performed on the second pass reverse osmosis adjust the operating pH and resulting caustic soda treatment rate.
Nisan, S., Dardour, S.	2007	Economic evaluation of nuclear desalination systems	Desalination 205 (2007) 231–242	This paper describes the detailed analyses of power and water costs for several nuclear reactors operating in a cogeneration mode and coupled to two main desalination processes, e.g. multiple effect distillation (MED) and reverse osmosis (RO).	France	NA	RO	WT/SW	Results based on the utilisation of this RO process, with the preheating of the feed water, called ROph, will also be presented later.
Dardour, S., Nisan, S., Charbit, F.	2007	Utilisation of waste heat from GT–MHR and PBMR reactors for nuclear desalination	Desalination 205 (2007) 254–268	This paper describes the successive stages that led to the development of physical and mathematical models enabling the calculation of desalination costs of the GT–MHR and the PBMR providing free heat.	France	NA	RO	WT/SW	Gas cooled high temperature reactors such as the GT–MHR and the PBMR are ideally suited for seawater desalination purposes since, by their very design, these reactors provide virtually free heat at appropriate temperatures for an MED desalination plant.
Daub, K., Emig, G., Chollier, M.J., Callant, M., Dittmeyer, R.	1999	Studies on the Use of Catalytic Membranes for Reduction of Nitrate in Drinking Water	Chemical Engineering Science 54 (1999) 1577–1582	Hydrogenation of nitrate using supported bimetallic Pd-Cu or Pd-Sn catalysts provides a promising new alternative for denitrification of	Germany	European Union	Catalytic Membranes	WT/DW	A detailed characterization of the membranes, e.g. in terms of the distribution of the active metals, by advanced spectroscopy and electron microscopy techniques seems to be

(*continued*)

Authors	Year	Title	Journal name, volume (issue)	Abstract	Country	Funding Agent	Technology used	USE (WT/WWT)	Recommendations
				drinking water with several advantages over established technologies such as reverse osmosis or biological denitrification.					necessary to define the targets for the preparation.
Stefan Holler, S., Trosch, W.	2001	Treatment of urban wastewater in a membrane bioreactor at high organic loading rates	Journal of Biotechnology 92 (2001) 95–101	In this paper, experimental data from a laboratory scale membrane bioreactor are presented.	Germany	NA	Membrane bioreactor	WWT/municipal	Additional work on the fouling behaviour of the microfiltration membrane as well as the oxygen supply of the reactor has to be made in order to get a further understanding of the processes in the membrane bioreactor. In subsequent studies a scale-up of the membrane bioreactor will be realized and the degradation of real wastewater will be investigated.
Chafik, E.	2002	A new seawater desalination process using solar energy	Desalination 153 (2002) 25–37	This paper presents the development of a new process to desalinate seawater using solar energy.	Germany	European Commission	Desalination	WT(SW&BW)	A new design of a low-cost collector for solar air heating has been developed. An efficient humidifier design is developed and tested.
Kleine, J., Peinemann, K.-V., Schuster, C., Warnecke, H.J.	2002	Multifunctional system for treatment of wastewaters from adhesive-producing industries: separation of solids and oxidation of dissolved pollutants using doted microfiltration membranes	Chemical Engineering Science 57 (2002) 1661–1664	To eliminate non-biodegradable organic compounds from wastewater application of semiconductor photocatalysis has been done.	Germany	Ministerium fur Umwelt Landwirtschaft und Verbraucherschutz des Landes Nordrhein-Westfalen.	MF	WWT/Industrial	A two-step process has been developed for the clarification of highly polluted waste waters from adhesive-producing plants. First, the suspended solids which reach up to 10% of the mass stream are precipitated, flocculated and separated by means of a decanting centrifuge and flotation. Then the photocatalytic process was applied on the dissolved organic contents.

(*continued*)

Authors	Year	Title	Journal name, volume (issue)	Abstract	Country	Funding Agent	Technology used	USE (WT/WWT)	Recommendations
Annaka, Y., Hamamoto, Y., Akatsu, M., Maruyama, K., Oota, S., Murakami, T.	2002	Development of MBR with reduced operational and maintenance costs	Water Science & Technology Vol. 53 No. 3 pp. 53–60	To reduce MBR O&M costs, a new MBR process that conducts efficient simultaneous biological nitrogen and phosphorus removal (BNR) was developed.	Germany	NA	MBR	WWT/Municipal	The new simultaneous biological nitrogen and phosphorus removal MBR process developed under this study and adopted for the demonstration experiment had a 6-h HRT and stably achieved excellent effluent characteristics.
Gehlert, G., Hapke, J.	2002	Mathematical modelling of a continuous aerobic membrane bioreactor for the treatment of different kinds of wastewater	Desalination 146 (2002) 405–412	In this paper, a mathematical model, which describes a continuous aerobic membrane bioreactor (MBR) for the treatment of different kinds of wastewater, is developed.	Germany	ROCHEM UF-SYSTEME GmbH and BEC GmbH	MBR, UF	WWT	Both models showed a reasonable agreement to experimental data obtained from experiments with size containing wastewater.
Issa, M., Robert, J., Denecke, M., Kummel, R.	2003	Water re-use in practice Closing loops in food industry, wastewater reduction in composting plants and municipal wastewater re-use for irrigation purposes	Management of Environmental Quality: An International Journal Vol. 14 No. 5, 2003 pp. 571–576	This paper shows, on the basis of three case studies, the possibilities of environmental protection on the one hand, and the achievement of financial benefits on the other hand by introducing membrane technology.	Germany	NA	UF	WWT/Municipal	Domestic wastewater can also be treated to be reusable for irrigation, which could help alleviate the discrepancy between diminishing water resources and growing demand for water from increasing industries such as agricultural.
Busch, M., Mickols, W.E.	2004	Reducing energy consumption in seawater desalination	Desalination 165 (2004) 299–312	This publication presents new products with higher flow (7500–9000gpd at standard test conditions) and high rejection (99.70–99.75%) for the seawater desalination segment.	Germany	OWV, DOW, INALSA, Delta, Caramondani	RO	WT/Sea water	Demonstrated significant enhancements in the rejection performance and the productivity of seawater reverse osmosis offer the potential of large reductions in water cost.

(*continued*)

Authors	Year	Title	Journal name, volume (issue)	Abstract	Country	Funding Agent	Technology used	USE (WT/WWT)	Recommendations
Blocher, C., Bunse, U., SeBler, B., Chmiel, H., Janke, H.D.	2004	Continuous regeneration of degreasing solutions from electroplating operations using a membrane bioreactor	Desalination 162 (2004) 315–326	The objective is to retain the biomass as well as the hydrocarbons in the bioreactor by the microfiltration membranes to enhance biodegradation. Simultaneously, low retention is aimed at for the surfactants since these substances have to be returned to the degreasing bath.	Germany	Germany Federal Ministry for Education and Research (BMBF)	MBR	WWT/Industrial	The tests described above will be continued in a test operation at industrial level which should provide information on the growth behaviour of biomass, the long-term performance of the process as well as a detailed economic analysis.
Seibert, U., Vogt, G.	2004	Autonomous desalination system concepts for sea water and brackish water in rural areas with renewable energies – ADIRA	International conference on desalination costing, Dec, 2004, Cyprus	ADIRA aims at the development of optimum concepts for fresh water supply in rural areas derived from salty water (sea water and brackish water).	Germany	EU	Desalination	WT/(Sea water & brackish water)	In order to gain a wide and profound knowledge of small-scale desalination units powered by renewable energies under real working conditions, it is essential to move out of the laboratories and to study the real projects in the field.
Lyko, S., Wintgens, T., Melin, T.	2005	Estrogenic trace contaminants in wastewater-possibilities of membrane bioreactor technology	Desalination 178 (2005) 95–105	In this study the possibilities of MBRs with regard to the elimination efficiency of estrogenic trace contaminants are considered for these two applications.	Germany	German Research Association (DFG)	MBR	WWT/Municipal	The characteristics of influents (municipal wastewater or landfill leachate) have an effect on the composition of the activated sludge and associated an impact on the possibilities of the membrane unit on micropollutant rejection.
Wintgens, T., Melin, T., Schiller, A., Khan, S., Muston, M., Bixio, D., Thoeye, C.	2005	The role of membrane processes in municipal wastewater reclamation and reuse	Desalination 178 (2005) 1–11	This paper will provide an overview of the status of membranes processes in wastewater reclamation and reuse world-wide and will	Germany	European Commission	RO&MF	WWT/Municipal	Future development will also have to include on-line quality control of membrane processes. As long as integrity failures can go undetected for longer times, relatively

(*continued*)

Authors	Year	Title	Journal name, volume (issue)	Abstract	Country	Funding Agent	Technology used	USE (WT/WWT)	Recommendations
				depict their potential role in promoting more sustainable water use patterns.					expensive dual or multiple barrier systems will be required for all those applications that require guaranteed drinking water quality at all times.
Panglisch, S., Holy, A., Urban, F., Dautzenberg, W., Sous, P., Ohligschlaigeff, J., Gimbel, R.	2005	Transferring pilot experiments into the planning of Germany's largest two-stage ultrafiltration plant	Desalination 179 (2005) 225–235	Numerous intensive preliminary investigations were carried out. Among other things the investigations showed that the membrane process will be most efficient if flocculated raw water is used.	Germany	NA	UF	WT/DW	In the future treatment process, chemical-free and acidic and caustic backwash water will accumulate separately.
Vocks, M., Adam, C., Lesjean, B., Gnirss, R., Kraume, M.	2005	Enhanced post-denitrification without addition of an external carbon source in MBR	Water Research 39 (2005) 3360–3368	This study investigates a post-denitrification process without the addition of an external carbon source combined with an enhanced biological phosphorus removal (EBPR) in a membrane bioreactor (MBR).	Germany	Berliner Wasserbetriebe and Veolia Water in the frame of the Kompetenzzentrum Wasser Berlin.	MBR	WWT/Municipal	The tests indicated that some glycogen accumulating bacteria (GAO) are probably responsible for the enhanced denitrification.
Melin,T., Jefferson, B., Bixio, D., Thoeye, C., De Wilde, W., De Koning, J., van der Graaf, J., Wintgens, T.	2006	Membrane bioreactor technology for wastewater treatment and reuse	Desalination 187 (2006) 271–282	This paper will provide an overview of the status of membrane bioreactor applications in municipal wastewater reclamation and reuse in Europe and will depict their potential role in promoting more sustainable water use patterns.	Germany	European Commission	MBR	WWT/Municipal	Many areas such as fouling control, pre-treatment, maintenance and operators training have to be established in the operational procedures and drawbacks are expected in the uptake of this technology, which define the need for more intensive and practitioner-oriented research on MBRs.
Broeckmann, A., Busch, J., Wintgens, T., Marquardt, W.	2006	Modelling of pore blocking and cake layer formation in membrane filtration	Desalination 189 (2006) 97–109	In this study, two phenomena are newly introduced into the existing models, which have a major impact	Germany	Thank Koch Membrane Systems GmbH	Hollow fiber membrane	WWT/Municipal	The additional models significantly increase the predictive capabilities of the filtration model and will be essential for designing

(*continued*)

Authors	Year	Title	Journal name, volume (issue)	Abstract	Country	Funding Agent	Technology used	USE (WT/WWT)	Recommendations
		for wastewater treatment		on cake layer and pore resistance. One phenomenon is the distribution of particle and membrane pore diameters and the other is the adhesion between particles and the membrane surface.					suitable operational strategies and process designs for membrane filtration applications.
García-Molina, V., Lyko, S., Esplugas, S., Wintgens, Th., Melin, Th.	2006	Ultrafiltration of aqueous solutions containing organic polymers	Desalination 189 (2006) 110–118	Ultrafiltration was the selected membrane process and membranes with different molecular weight cut off (MWCO) have been employed. The feed of the experiments consisted of solutions containing humic and fulvic acids, natural organic matter (NOM), alginic acid and cellulose.	Germany	NA	UF	WWT/Industrial	Increasing transmembrane pressure is related to an increase in flux and a decrease in permeability. The calcium content of the feed was found to be influencing the flux when working with NOM and alginic acid, since lower fluxes were observed when adding Ca^{2+} to the feed.
Cornel, P., Krause, S.	2006	Membrane bioreactors in industrial wastewater treatment – European experiences, examples and trends	Water Science & Technology Vol. 53 No. 3 pp. 37–44	The aim of the paper is to share the European experiences with the MBR process, to elaborate the differences to the conventional activated sludge process and to present a possible technical solution for the design of the MBR process.	Germany	NA	MBR	WWT/Industrial	The further development and research of membrane bioreactors will be in: the improvement of cleaning strategies, especially in the application of safe chemicals for the water body the optimisation of the energy demand, improvement of the fouling control (maximum utilisation of the used air) membranes longer lifetime high and stable flux and high permeability low cost resistant against mechanical forces.

(*continued*)

Authors	Year	Title	Journal name, volume (issue)	Abstract	Country	Funding Agent	Technology used	USE (WT/WWT)	Recommendations
Baumgarten, S., Schröder, H.Fr., Pinnekamp, J.	2006	Performance of MBRs used for the treatment of wastewater from the chemical and textile industries	Water Science & Technology Vol. 53 No. 3 pp. 61	Within the scope of the study, nine waste waters from the chemical and textile industries were treated in bench-scale (laboratory scale) and small-scale (pilot scale) membrane bioreactors.	Germany	NA	MBR	WWT/Municipal	Since the MBR filtrates are free from particulate substances and have a low fouling potential, advantages compared to pre-treatment in conventional systems can be expected.
Melin,T., Jefferson, B., Bixio, D., Thoeye,C., De Wilde,W., De Koning, J., van der Graaf, J., Wintgens, T.	2006	Membrane bioreactor technology for wastewater treatment and reuse	Desalination 187 (2006) 271–282	This paper will provide an overview of the status of membrane bioreactor applications in municipal wastewater reclamation and reuse in Europe and will depict their potential role in promoting more sustainable water use patterns.	Germany	EU	MBR	WWT/Municipal	The operational experience with full-scale MBRs indicates that this relatively new technology poses a challenge to the water utilities. Many areas such as fouling control, pre-treatment, maintenance, and operators training have to be established in the operational procedures and drawbacks are expected in the uptake of this technology, which define the need for more intensive and practitioner-oriented research on MBRs.
Blank, J.E., Tusel, G.F., Nisan, S.	2007	The real cost of desalted water and how to reduce it further	Desalination 205 (2007) 298–311	In this context, the majority of the media are underlining the importance of seawater desalination as an attractive and logical alternative source to fight the freshwater scarcity.	Germany	NA	Desalination	WT/SW	The basic conclusion of our investigation is that large thermal and membrane desalination plants may reach in the future the same level of specific water cost and that a water production cost of 1.00€/m^3, with energy cost of the year 2006 can be achieved with moderate R&D efforts. However, it is difficult (not impossible) to reach 0.50€/m^3 and it will require time-consuming and intensive R&D work to achieve this goal.

(*continued*)

Authors	Year	Title	Journal name, volume (issue)	Abstract	Country	Funding Agent	Technology used	USE (WT/WWT)	Recommendations
Deshmukh, S.A. R.K., Heinrich, S., Mörl, L., Annaland, M.V.S., Kuipers, J.A.M.	2007	Membrane assisted fluidized bed reactors: Potentials and hurdles	Chemical Engineering Science 62 (2007) 416–436	This paper seeks to review literature and new developments in the area of membrane assisted fluidized bed reactors, with special emphasis on possible applications, integration of different membranes in the fluidized beds, reactor modelling studies, experimental demonstration of various reactor concepts and future prospects and hurdles for commercialization.	Germany	NA	MAFBR	WT	The progress that has been made till now suggests that the challenges to the commercialization of the MAFBRs will be met in the future, especially for controlled oxygen/air dosing using porous membranes for partial oxidation reactions and perm-selective hydrogen withdrawal in steam or autothermal steam reforming for ultra-pure hydrogen production.
Ernst, M., Sperlich, A., Zheng, X., Gan, Y., Hu,J., Zhao, X., Wang, J., Jekel, M.	2007	An integrated wastewater treatment and reuse concept for the Olympic Park 2008, Beijing	Desalination 202 (2007) 293–301	The present study summarizes preliminary results of the Sino-German research cooperation project on Sustainable Water Concept and its Application for the Olympic Games 2008 in the subproject 'Wastewater Treatment and Reuse'.	Germany	German Ministry of Education and Research and the Chinese Research Ministry (MOST).	Membrane bioreactors, specific phosphorus (P) adsorption columns and ultrafiltration (UF) with nature-based treatment processes	WWT/Municipal	In laboratory scale column tests with real MBR effluents, GFH could be repeatedly regenerated by caustic solutions with an efficiency of 50%. According to the consortiums preliminary results, the existing water reuse regulation for nutrients (TP and TN) does not seem to be sufficient if treated municipal effluents are to be used as the only source for an artificial lake in the Olympic Green. Further pilot-scale tests are in progress.
Manolakos, D., Papadakis, G., Sh. Mohamed, E., Kyritsisa, S., Bouzianas, K.	2005	Design of an autonomous low-temperature solar Rankine cycle system for reverse osmosis desalination	Desalination 183 (2005) 73–80	The present paper regards the design outline, of a low temperature solar organic Rankine cycle system for reverse osmosis (RO) desalination.	Greece	European Commission.	RO	WT/SW	The higher the temperature difference between high and low temperature reservoir, the higher the efficiency. Butthescopeof research is to examine the behaviour of Rankine system at

(*continued*)

Authors	Year	Title	Journal name, volume (issue)	Abstract	Country	Funding Agent	Technology used	USE (WT/WWT)	Recommendations
									temperatures in the range of 70–80°C, so as to extract tangible results about the exploitation of low temperature thermal sources.
Sh. Mohamed, E., Papadakis, G., Mathioulakis, E., Belessiotis, V.	2005	The effect of hydraulic energy recovery in a small sea water reverse osmosis desalination system; experimental and economical evaluation	Desalination 184 (2005) 241–246	This paper presents the experimental results of a small Sea Water Reverse Osmosis desalination system equipped with a hydraulic energy recovery unit of the Eco systems Clark pump type.	Greece	European Commission	RO	WT/SW	An energy and cost efficient small desalination system is installed and tested.
Yiantsios, S.G., Sioutopoulos, D., Karabelas, A.J.	2005	Colloidal fouling of RO membranes: an overview of key issues and efforts to develop improved prediction techniques	Desalination 183 (2005) 257–272	In the present RO fouling study, humic acid was selected as a representative contaminant for organic fouling.	Greece	Middle East Desalination Research Centre	RO	WWT/Industrial	There certainly exists the need for improved means of predicting the colloidal fouling propensity of feed waters, in the direction of determining actual flux decline and accounting for membrane type and operating conditions. Important considerations in pursuing improvements in fouling indices is to simulate as closely as possible the conditions prevailing in actual RO membrane modules, namely flux, concentration polarization and membrane/fouling interactions, and to predict fouling rates.
Pagana, A., Stoitsas, K., Zaspalis, V.T.	2006	Applied pilot-scale studies on ceramic membrane processes for the treatment of wastewater streams.	Global NEST Journal, Vol. 8, No. 1, pp. 23–30, 2006	In this article the results of the large area synthesis and pilot-scale experiments for porous ceramic membranes are reported for three case studies: I) purification	Greece	Greek Ministry of Development	Ceramic membranes	WT&WWT (Olive Oil)	Complete microparticle rejection of wastewater streams at an average solids free water production of $0.8 m^3 hr^{-1} m^{-2}$ with operation under an average pressure difference of $3 \times 10^5 Nt\, m^{-2}$.

(*continued*)

Authors	Year	Title	Journal name, volume (issue)	Abstract	Country	Funding Agent	Technology used	USE (WT/WWT)	Recommendations
				of water from suspended microparticles (i.e. d50~0.5μm), II) purification of water from suspended nanoparticles (i.e. d50~30nm) and III) treatment of olive oil refinery wastewater.					• Complete nanoparticle rejection of wastewater streams at an average solids free water production of 0.16m^3 hr^{-1} m^{-2} with operation under an average pressure difference of 3×10^5Nt m^{-2}. Olive oil mill wastewater can be upgraded with reduction of its BOD5 content by 61% and its concentration in phenols by 82% at an average dischargeable waste production of 0.01m^3 hr^{-1} m^{-2} with operation under an average pressure difference of 3×10^5Nt m^{-2}.
Criscuoli, A., Drioli, E.	1999	Energetic and exergetic analysis of an integrated membrane desalination system	Desalination 124 (1999) 243–249	An integrated reverse osmosis (RO) + membrane distillation (MD) system where MD operates on the RO brine is considered for desalination processes	Italy	NA	RO&MD	WT/Sea water	The integrated MD system has a better performance with respect to the thermal processes as MSF and might represent an interesting option for desalination operations.
Drioli, E., Laganh, F., Crlscuoh, A., Barbieri, G.	1999	Integrated membrane operations in desalination processes	Desalination 122 (1999) 141–145	In the present work, an integrated desalination membrane system for a high recovery factor based on the use of MD operating on the brine of a RO unit is presented and discussed. The possibility of introducing several	Italy	NA	RO	WT/Sea water	Integrating membrane operations it is possible to improve the performance of a seawater desalination system with a higher recovery factor with respect to an RO unit alone by maintaining acceptable costs. An intelligent combination of several membrane units

(*continued*)

Authors	Year	Title	Journal name, volume (issue)	Abstract	Country	Funding Agent	Technology used	USE (WT/WWT)	Recommendations
				membrane operations for pretreatment has also been considered.					will lead to the rationalization of industrial production for the recovery and reuse of water, with a reduction of environmental problems and an optimization of water consumption.
Giorno, L., Drioli, E.	2000	Biocatalytic membrane reactors: applications and perspectives	Tibtech august 2000 (vol.18) page 339–349	Membranes and biotechnological tools can be used for improving traditional production systems to maintain the sustainable growth of society. Typical examples include: new and improved foodstuffs, in which the desired nutrients are not lost during thermal treatment; novel pharmaceutical products with well-defined enantiomeric compositions; and the treatment of wastewater, wherein pollution by traditional processes is a problem.	Italy	NA	Biocatalytic membrane reactors	Pharmaceuticals, biomedical applications, wastewater treatment	The potential advantages of membrane reactor technology over more conventional approaches include its higher efficiency and reduced costs owing to the integration of bioconversion and product purification, thus reducing equipment costs and the number of processing steps.
Criscuoli, A., Basile, A., Drioli, E.	2000	An analysis of the performance of membrane reactors for the water–gas shift reaction using gas feed mixtures	Catalysis Today 56 (2000) 53–64	The objective of this work has been to analyse the effect of the feed composition on the membrane reactor efficiency in order to determine the best conditions in terms of CO conversion. At this purpose, experimental tests with mixtures of different compositions have been carried out in three different	Italy	Authors are grateful to Dr. Violante and Dr. Tosti (Enea, Rome, Italy) for the palladium reactor supply and to Dr. P. Thygesen and Dr. N. Nielsen (Haldor Topsoe A/S, Copenhagen, Denmark) for providing the catalyst for the water–gas shift reaction.	Catalytic membrane reactors	Water–gas shift reaction	The better performance of the palladium membrane with respect to the ceramic one is due not only to the selectivity but also to the hydrogen permeation rate. The fact that different kinetic expressions have to be used to describe the fixed-bed reactor and membrane reactors confirms that in membrane reactors the kinetic mechanism might be

(*continued*)

Authors	Year	Title	Journal name, volume (issue)	Abstract	Country	Funding Agent	Technology used	USE (WT/WWT)	Recommendations
				systems of reaction: (1) traditional fixed-bed reactor; (2) membrane reactor with mesoporous ceramic membrane; (3) membrane reactor with palladium membrane.					different than in a traditional system.
Molinari, R., Mungari, M., Drioli, E., Di Paola, A., Loddo, V., Palmisano, L., Schiavello, M.	2000	Study on a photocatalytic membrane reactor for water purification	Catalysis Today 55 (2000) 71–78	The behaviour of the membrane photoreactor using the two different confination techniques of TiO_2 was studied. 4-NP was chosen as a model molecule because its photodegradation in aqueous TiO_2 suspensions, irradiated with UV–VIS light, was already studied.	Italy	IRMERC-CNR (CS) and the Ministero dell'Università e della Ricerca Scientifica e Tecnologica (Rome)	Photo-catalytic membrane	WT	Polymeric membranes can be used for photocatalytic reactions, but it is necessary to carry out preliminary tests to choose suitable polymers stable under UV–VIS illumination.
Szpyrkowicz, L., Kelsall, G.H., Kaul, S.N., Faveri, M.D.	2001	Performance of electrochemical reactor for treatment of tannery wastewaters	Chemical Engineering Science 56 (2001) 1579–1586	An undivided electrochemical reactor with parallel plate Ti/Pt its anode and stainless-steel cathode was used to treat tannery wastewater at constant current densities and different stirring rates.	Italy	CNR (Italy) and CSIR (India)	Electro-chemical reactor	WWT/tannery	The feasibility has been demonstrated of oxidizing the principle components: ammonium ions, sulphate, tannins and other organics (TOC and COD), of real and synthetic tannery wastewaters by ectrochemically generated active chlorine, formed by chloride oxidation in an undivided electrochemical batch reactor with parallel plate electrodes.
Criscuoli, A., Basile, A., Drioli, E., Loiacono, O.	2001	An economic feasibility study for water gas shift membrane reactor	Journal of Membrane Science 181 (2001) 21–27	An economical analysis of palladium membrane reactors for the water gas shift (WGS) reaction is presented and discussed.	Italy	NA	Palladium membrane reactors	Water gas	The rationalisation of various chemical processes might be possible by integrating membrane reactors in the whole system of industrial chemical productions.

(continued)

Authors	Year	Title	Journal name, volume (issue)	Abstract	Country	Funding Agent	Technology used	USE (WT/WWT)	Recommendations
Curcio, E., Criscuoli, A., Drioli, E.	2001	Membrane Crystallizers	Ind. Eng. Chem. Res. 2001, 40, 2679–2684	The purpose of this work is to introduce an innovative methodology, the membrane crystallization, to produce crystals from solutions.	Italy	NA	MD	Aqueous solutions	The achieved results, in substantial agreement with those reported in the literature, confirm the interesting potentialities of the membrane crystallization strategy in redesigning traditional crystallizers.
Molinari, R., Grande, C., Drioli, E., Palmisano, L., Schiavello, M.	2001	Photocatalytic membrane reactors for degradation of organic pollutants in water	Catalysis Today 67 (2001) 273–279	In this paper, some experimental results on a batch photocatalytic reactor without membrane to find the influence of the characteristic variables of the process and a kinetic equation for degradation rate using a semi-empirical approach are reported.	Italy	IRMERC-CNR (CS) and the MURST (Rome) (Progetto L.488 of INCA Consortium)	Photo-catalytic membrane	WT/4-Nitrophenol (4NP)	These results suggest the use of the hybrid continuous membrane photoreactor as a new scheme for some industrial applications of the photocatalytic technique. Further improvements of this process are under study.
Cassano, A., Molinari, R., Romano, M., Drioli, E.	2001	Treatment of aqueous effluents of the leather industry by membrane processes A review	Journal of Membrane Science 181 (2001) 111–126	A general overview on the potentiality of membrane processes in the treatment of aqueous solutions coming from the leather industry is reported.	Italy	NA	UF, NF&RO	WT/leather industry	The described applications confirm the new important possibilities of having this industrial sector a sustainable growth also in industrialized countries.
Marcucci, M., Nosenzo, G., Capannelli, G., Ciabatti, I., Corrieri, D., Ciardelli, G.	2001	Treatment and reuse of textile effluents based on new ultrafiltration and other membrane technologies	Desalination 138 (2001) 75–82	Different membrane processes were experimented on at pilot scale to verify the possibility of reusing textile wastewater.	Italy	Regione Toscana	UF, NF&RO	Textile wastewater	Considering the positive results of the pilot scale experimental studies, the implementation of the technique on the industrial scale is considered technically and economically feasible, and it could reduce the freshwater drain and the wastewater discharge by half.

(continued)

Authors	Year	Title	Journal name, volume (issue)	Abstract	Country	Funding Agent	Technology used	USE (WT/WWT)	Recommendations
Driolli, E., Criscuoli, A., Curcio, E.	2002	Integrated membrane operations for seawater desalination	Desalination 147 (2002) 77–81	Reverse osmosis is widely applied to water desalination because of the relatively low cost of the process.	Italy	NA	RO	WT/Sea water	Adoption of integrated membrane systems appears an interesting possibility for improving desalination operations and meeting the increasing pure water demand.
Cassano, A., Molinari, J.A.R., Buonomenn, M.G., Roig, J., Drioli, E.	2003	Membrane treatment by nanofiltration of exhausted vegetable tannin liquors from the leather industry	Water Research 37 (2003) 2426–2434	A nanofiltration (NF) process was used for recovering tannins and water from exhausted baths and reusing them as tanning agents and washings.	Italy	European Union	NF	WWT/Industrial	The advantages are particularly significant in terms of reduction of environmental impact (up to 75% of COD reduction of the global effluent), simplification of cleaning-up processes of wastewaters, saving of chemicals and water, easy reuse of sludge and decrease of disposal costs.
Pollice, A., Laera, G., Blonda, M.	2004	Biomass growth and activity in a membrane bioreactor with complete sludge retention	Water Research 38 (2004) 1799–1808	This work reports the main results of a bench scale membrane bioreactor operated for more than 100 days without sludge withdrawal and fed on real municipal wastewater.	Italy	Zenon Environmental S.r.l.	MBR	WWT/Municipal	The evaluation of the need and frequency of sludge withdrawal for limiting build-up of inert materials within the plant will have to be assessed over the long term.
Van der Bruggen, B., Johannes, C., Jansen, J., Figoli, A., Geens, J., Van Baelen, D. Drioli, E., Vandecasteele, C.	2004	Determination of Parameters Affecting Transport in Polymeric Membranes: Parallels between Pervaporation and Nanofiltration	J. Phys. Chem. B 2004, 108, 13273–13279	The differences in structure and transport mechanisms of nanofiltration (NF) and pervaporation (PV) membranes were studied by measuring (a) water and ethanol fluxes and raffinose rejections; (b) water and ethanol fluxes and water/ethanol-n-propanol separation factors; (c) gas and vapour (water, ethanol).	Italy	FWOVlaanderen postdoctoral grant for B. Van der Bruggen.	NF&PV	Water & ethanol	Changes of the nature of the membrane under different conditions offer the perspective of using the same membrane for different types of separation processes.

(*continued*)

Authors	Year	Title	Journal name, volume (issue)	Abstract	Country	Funding Agent	Technology used	USE (WT/WWT)	Recommendations
Molinari, R., Buonomenn, M.G., Cassano, A., Drioli, E.	2004	Recovery and recycle of tannins in the leather industry by nanofiltration membranes	Chem Technol Biotechnol 79:361–368 (online: 2004)	The aim of this work was the study of the application of membrane concentration to exhausted tanning baths from vegetable tannage operations in order to increase the tannin/non-tannin (T/NT) ratio and obtain recyclable material.	Italy	NA	NF	WWT/Industrial	The study of solute–membrane interaction explained the better performance of the TFC S membrane. The lesser extent of hydrogen bonding between tannins and the TFC S membrane can explain the lesser fouling tendency of this membrane with respect to the NF 45 membrane.
Drioli, E., Curcio, E., Criscuoli, A., Di Profio, G.	2004	Integrated system for recovery of $CaCO_3$, NaCl and $MgSO_4.7H_2O$ from nanofiltration retentate	Journal of Membrane Science 239 (2004) 27–38	An integrated membrane system has been developed in order to recover dissolved salts that are present, at low concentration, in typical feed streams to desalination plants.	Italy	NA	NF	WT/Sea water	The possibility (if the retentate stream is entirely processed) to increase the fresh water recovery factor of the nanofiltration unit up to 95%, to recover 78% of dissolved NaCl, and to produce 8.4kg $SO_4.7H_2Ocm^{-3}$ of NF retentate has been experimentally verified.
Cassano, A., Jiao, B., Drioli, E.	2004	Production of concentrated kiwifruit juice by integrated membrane process	Food Research International 37 (2004) 139–148	This paper describes the research efforts to develop and optimise an integrated membrane process, on laboratory scale, for the production of concentrated kiwifruit juice as alternative to the traditional vacuum evaporation.	Italy	Prof. Rosangela Marchelli and Dr. Gianni Galaverna of the Department of Organic and Industrial Chemistry of University of Parma	UF & OD	Kiwifruit juice	According to the results obtained, an integrated membrane process for the production of concentrated kiwifruit juice of high quality and high nutritional value was proposed.
Pizzichini, M., Russo, C., Di Meo, C.	2005	Purification of pulp and paper wastewater, with membrane technology, for water reuse in a closed loop	Desalination 178 (2005) 351–359	The wastewaters of an Italian paper mill (Lucart), employing recycled paper, news print and cardboard as raw materials, were	Italy	NA	MF, UF, NF &RO	WWT/Industrial	The best filtration performance was obtained with the ceramic tubular membrane, which gives higher productivities and lower fouling.

(*continued*)

Authors	Year	Title	Journal name, volume (issue)	Abstract	Country	Funding Agent	Technology used	USE (WT/WWT)	Recommendations
				purified with membrane technologies with the aim to reuse the purified water in a closed loop in the paper manufacturing process.					
Maragliano, G., Moss, P.	2005	The development of a high flow seawater membrane. A case history of one of the first applications using high flow seawater elements in a plant producing process and boiler feed water for ENEL (now EDIPOWER) at San Filippo del Mela power plant in Italy	Desalination 184 (2005) 247–252	This paper is a case history describing the plant and looking at the performance of the membrane using Koch Membrane Systems NormPro1 to normalise the data.	Italy	NA	RO	WT/Sea water	The first full commercial production of the Fluid SystemsTM TFC1 2822HF-370 elements performed very well on start up in 1999 and has continued to perform well over the 4 years to date. The client has been very satisfied with the performance.
Laera, G., Pollice, A., Saturno, D., Giordano, C., Lopez, A.	2005	Zero net growth in a membrane bioreactor with complete sludge retention	Water Research 39 (2005) 5241–5249	A bench-scale membrane bioreactor was operated with complete sludge retention in order to evaluate biological processes and biomass characteristics over the long term.	Italy	NA	MBR	WWT/Municipal	Larger scale studies are needed to evaluate all corelated issues, including aeration, membrane cleaning and sludge treatability.
Giorno, L., Mazzei, R., Oriolo, M., De Luca, G., Davoli, M., Drioli, E.	2005	Effects of organic solvents on ultrafiltration polyamide membranes for the preparation of oil-in-water emulsions	Journal of Colloid and Interface Science 287 (2005) 612–623	Hydrophilic ultrafiltration membranes made of polyamide with molecular weight cut-off 10 and 50kDa have been studied for the preparation of oil-in-water emulsions by a cross-flow membrane emulsification technique.	Italy	Ministero degli Affari Esteri, Direzione Generale per la Promozione e la Cooperazione Culturale.	UF	O/W emulsion	The pretreatment methodology made it possible to obtain very high fluxes with low transmembrane pressure for a direct oil-in-water emulsification.

(*continued*)

Authors	Year	Title	Journal name, volume (issue)	Abstract	Country	Funding Agent	Technology used	USE (WT/WWT)	Recommendations
Drioli, E., Di Profio, G., Curcio, E.	2005	Hybrid membrane operations in water desalination and industrial process rationalisation	Water Science & Technology Vol 51 No 6–7 pp 293–304	In the present paper, other than for seawater desalination applications, some industrial applications where membrane technology has led already to match the goal of process intensification are discussed.	Italy	NA	RO&MSF	WT/Sea water	If process intensification is the most promising opportunity for a sustainable industrial development, membrane technology agrees perfectly in this respect.
Zhang, J., Giorno, L., Drioli, E.	2006	Study of a hybrid process combining PACs and membrane operations for antibiotic wastewater treatment	Desalination 194 (2006) 101–107	The objective of this study was to find a suitable method to purify wastewater coming from the production and downstream processing of TC.	Italy	NA	RO&UF	WWT/Antibiotic (Industrial)	Based on this study, the integrated PACs–UF–RO process can be considered a promising method to treat TC wastewater.
Lopez, A. Pollice, A, Lonigro, A., Masic, S., Palesed, A.M., Cirellie, G.L., Toscanoe, A., Passino, R.	2006	Agricultural wastewater reuse in southern Italy	Desalination 187 (2006) 323–334	The efficiencies of four technological options (membrane filtration, simplified treatments, wastewater storage reservoirs, constructed wetlands) are evaluated, at demonstrative scale, as viable alternatives for reusing reclaimed municipal effluents in agriculture.	Italy	NA	Membrane filtration	WWT/Municipal	Membrane filtration. The microbial quality of treated effluents was higher than that of the benchmark (i.e., local well-water conventionally used for irrigation). Storage reservoirs. After appropriate storage periods: TSS, BOD5, COD and nutrients concentrations achieved the in force Italian limits for WW agricultural reuse. Constructed wetlands. Recorded average efficiencies for TSS, BOD5, COD, TN and TP removal resulted 85%, 65%, 75%, 42% and 32% respectively. Simplified treatments. Effluents obtained by skipping biological processes for saving the

(*continued*)

Authors	Year	Title	Journal name, volume (issue)	Abstract	Country	Funding Agent	Technology used	USE (WT/WWT)	Recommendations
									agronomic potential of organic matter and nutrients occurring in urban wastewater, once tertiary treated, had a microbiological quality not significantly worse than that of conventional secondary effluents also tertiary treated.
Brunetti, A., Barbieri, G., Drioli, E.	2006	A PEMFC and H_2 membrane purification integrated plant	Desalination 199 (2006) 156–158	In this work particular attention was focused on the integration between a Pd-based membrane purification step and a commercial PEMFC.	Italy	RENCO S.p.A (Italy)	Pd-based membrane	H_2	No performance loss is registered when the PEMFC was fed from membrane purification step confirming that the Pd-based membrane purificator assures a CO free H_2 stream.
Chiappetta, G., Clarizia, G., Drioli, E.	2006	Design of an integrated membrane system for a high level hydrogen purification	Chemical Engineering Journal 124 (2006) 29–40	Aim of this work is the theoretical analysis of the potential of integrated membrane systems to recover hydrogen at a very high purity level (with CO content lower than 10ppm), suitable for fuel cell applications.	Italy	NA	Polymeric and palladium (Pd) separators as well as a Pd-based membrane	Hydrogen purification	In this work the potential of different integrated membrane systems has been evaluated to recover hydrogen (90% of the feed value) at a high purity level (99.9995 mol%) and reduce as possible the CO content in the retentate stream, by controlling the expensive Pd membrane surface and compression power.
Chiappetta, G., Clarizia, G., Drioli, E.	2006	Analysis of safety aspects in a membrane reactor	Desalination 193 (2006) 267–279	Membrane reactors can combine reaction and separation steps in a single unit, improving the efficiency of operations.	Italy	NA	Non-isothermal membrane reactor	Water gas shift reaction	This aspect can be emphasised in the case of highly exothermic systems where the choice of a proper membrane material can favour the heat removal with the product permeation or avoid, by means of a controlled feed of the reagents, operations inside the flammability region.

(*continued*)

Authors	Year	Title	Journal name, volume (issue)	Abstract	Country	Funding Agent	Technology used	USE (WT/WWT)	Recommendations
Mariah, L., Buckley, C.A., Brouckaert, C.J., Curcio, E., Drioli, E., Jaganyi, D., Ramjugernath,D.	2006	Membrane distillation of concentrated brines – Role of water activities in the evaluation of driving force	Journal of Membrane Science 280 (2006) 937–947	Membrane distillation crystallisation (MDC) can be used to recover crystalline products from solutions. MDC of a concentrated solution of magnesium sulphate of 375g/l was investigated.	Italy	NA	Membrane distillation crystallisation (MDC)	Sasol (Pty) Ltd., the National Research Foundation (NRF) and the Technology and Human Resources for Industry Programme (THRIP)	These need to be well understood as the process must avoid crystallisation in the membrane module while producing a high quality final product which influences the performance of downstream processes (solid–liquid separation steps, drying and storage).
Cardona, E., Piacentino, A., Marchese, F.	2007	Performance evaluation of CHP hybrid seawater desalination plants	Desalination 205 (2007) 1–14	In this paper a small size (2000m^3/day) thermal desalination system (MEE) is coupled with a single-stage seawater reverse osmosis (SWRO) system; the system is fed by a natural gas (NG) reciprocate engine, where heat is recovered both from exhaust gases and from the cooling jacket water circuit.	Italy	NA	RO/MSF	WT/SW	Even being the system significantly unbalanced from an energetic point of view, when it is connected in parallel to the grid to sell surplus electricity production the obtained results suggest small-scale CHP integrated with hybrid MEE-RO systems to represent a viable solution.
Shatanawi, M. and Fayyad, M.	1996	Effect of Khirbet As-Samra treated effluent on the quality of irrigation water in the Central Jordan Valley	Desalination 1996, Pages 2915–2920, Vol. 30, Issue 12	The treated water has adversely affected the water quality of Yarmouk River. Comparison of the results of water quality tests with guidelines for water to be used for irrigation.	Jordan	NA		WWT/domestic	Salt tolerance of agricultural crops and the influence of water quality on the potential clogging of drip irrigation systems reveal that treated effluent from KA can be used for irrigation with restrictions. Concentrations of all trace elements were found to be low and within guidelines for irrigation water.
Akash, B.A., Al-Jayyousi, O.R., Mohsen, M.S.	1997	Multi-criteria analysis of non-conventional energy technologies for water desalination in Jordan	Desalination 114 (1997) 1–12	In this paper non-conventional energy technologies for water desalination are discussed.	Jordan	NA	RO	WT/BW	Further research and studies in this area are needed to ensure technical and economic feasibility of each technology.

(*continued*)

Authors	Year	Title	Journal name, volume (issue)	Abstract	Country	Funding Agent	Technology used	USE (WT/WWT)	Recommendations
Mohsena, M.S., Al-Jayyousib, O.R.	1999	Brackish water desalination: an alternative for water supply enhancement in Jordan	Desalination 124 (1999) 163–174	Brackish waters in Jordan are viewed as a potential and viable resource to alleviate water scarcity and overcome water budget deficit. The evaluation of various desalination technologies was carried out using multi-criteria analysis.	Jordan	NA	RO, electodialysis (ED), MSF, MED & VC	WT(brackish)	When evaluating various desalination processes in Jordan, RO was ranked the highest. The criteria under which multicriteria analyses were performed were economic, technical, and environmental.
Al-Hadidi, M.S.	1999	Brackish water management and use in Jordan	Desalination 126 (1999) 41–44	Desalination of brackish water seems to offer a reasonable source of supply augmentation as a non-conventional water resource.	Jordan	MOWI, MOA, MOH, MOI, MOT, HCST, RSS, PS and NGOs.	MSF,RO&ED.	WT(brackish)	It is recommended that an assessment of brackish water resources in terms of quantity, quality, and location for all basins should be carried out.
Al-Jayyousia, O.R., Mohsen, M.S.	2001	Evaluation of small home-use reverse osmosis units in Jordan	Desalination 139 (2001) 237–247	This paper aims to asses the quality of treated water from domestic RO units used in Jordan and to compare it with tap and bottled water in Amman.	Jordan	NA	RO	WT/fresh	In terms of quality and cost, it is found that RO produced water N unit provides water within the allowable limits of the Jordanian standards, with a relatively affordable reasonable price.
Mohsen, M.S., Jaber, J.O.	2001	A photovoltaic-powered system for water desalination	Desalination 138 (2001) 129–136	A reverse osmosis (RO) desalination system driven by photovoltaic power is proposed.	Jordan	NA	RO	WT/(seawater and brackish water)	Experimental work needs to be carried out to study the performance of this system and economic assessment of the proposed system should be performed.
AL-Jayyousi, O.	2001	Capacity building for desalination in Jordan: necessary conditions for sustainable water management	Desalination 141 (2001) 169–179	This paper highlights the Significance of water institutions in CB. It also aims to distinguish between water 'wants' and 'needs'. Also, it intends to address the issue of appropriate	Jordan	NA	RO	WT/(SW&BW)	Should be strengthened Partnership programs between Donor countries and Universities, Demand for desalinated water must be developed and enhanced. However, special attention must be

(*continued*)

Authors	Year	Title	Journal name, volume (issue)	Abstract	Country	Funding Agent	Technology used	USE (WT/WWT)	Recommendations
				technology and to introduce the concept of 'social adaptive capacity' to deal with water crisis in Jordan.					given to serving poor and rural people.
Jaber, J.O., Mohsen, M.S.	2001	Evaluation of non-conventional water resources supply in Jordan	Desalination 136 (2001) 83–92	This paper describes the development of a decision-support system for the evaluation and selection of potential non-conventional water resources supply; these include desalination of brackish and seawater, treated waste water, importation of water across boundaries and water harvesting.	Jordan	NA	Desalination	WT&WWT (BW, SW, domestic)	Further studies and training need to be carried out in the near future in order to promote and enforce the technology transfer in the field of desalination among the interested groups, especially governmental institutions, in Jordan.
Mohsen, M.S., Jaber, J.O.	2002	Potential of industrial wastewater reuse	Desalination 152 (2002) 281–289	The objective of this work is to evaluate the potential of industrial wastewater reuse in Jordan. Industrial water requirements, wastewater production, types of pollutants in industrial wastewater and the technologies for wastewater treatment have been discussed.	Jordan	NA	Desalination	WWT/Industrial	Evaluation of the need for upgrading the existing or building new wastewater treatment plants at the selected industries will be carried out.
Dweiri, S.F., Badran, M.I.	2002	Desalination: an imminent solution for the future water needs in the Aqaba Special Economic Zone (ASEZ)	Desalination 152 (2002) 27–39	This paper discusses the expected future water demands, the ASEZA acceptable water standards for the different water uses, and the environmental impact of the desalination plants and their respective mitigation measures.	Jordan	NA	RO	WT/(seawater and brackish water)	It is required from the desalination plants operators to provide mitigation measures for every component that impacts the environment and to demonstrate how these will lessen the damage to the environment.

(*continued*)

Authors	Year	Title	Journal name, volume (issue)	Abstract	Country	Funding Agent	Technology used	USE (WT/WWT)	Recommendations
Dababneh A. J., Al-Nirn, M.A.	2002	A reverse osmosis desalination unit	Desalination 153 (2002) 265–272	RO desalination unit is proposed to desalinate seawater. The pressure required is provided by utilizing the mechanical potential energy results from the differences in head between a high level column of seawater and a low level column of purified water.	Jordan	NA	RO	WT/Sea water	It is found that the specific gravity ratio has insignificant effect on the unit production; the unit production rate is linearly proportional to the saline water head Hand as the height of the purified water head rate decreases.
Abu Qdais, H.A., Batayneh, F.	2002	The role of desalination in bridging the water gap in Jordan	Desalination 150 (2002) 99–106	Review the status of water resources of Jordan and to evaluate the role of water desalination option in affecting the water supply equation of the country.	Jordan	NA	RO	WT/(seawater and brackish water)	Seawater desalination appears to be feasible for industrial and tourist sectors along the Red Sea shoreline. It is necessary to start programs for capacity building. To achieve this, it is required to create a basis for cooperation between the academic and research communities with the MWI.
Afonso, M.D., Maria, A., Alves, B., Mohsen, M.	2002	Crossflow microfiltration of marble processing wastewaters	Desalination 149 (2002) 153–162	MF is an efficient and ecologically suited environmental technology for decontamination and recycling of the wastewaters generated by marble processing plants, besides the environmental pollution abatement involved.	Jordan	NA	MF	WWT/Domestic	Continuous experiments should be carried out at higher concentration factors and longer operation periods, in order to get better insight of the concentration effect upon the concentration polarisation phenomena and/or cake build-up and thereby optimise the water recovery rate during microfiltration of marble wastewaters.
Banat, F., Jumah, R., Garaibeh, M.	2002	Exploitation of solar energy collected by solar stills for	Renewable Energy 25 (2002) 293–305	The aim of this work was to evaluate the technical feasibility of producing potable	Jordan	NA	ED, MD	WT/Sea water	The brine flow rate significantly affected the flux of the membrane module. The solar still was

(*continued*)

Authors	Year	Title	Journal name, volume (issue)	Abstract	Country	Funding Agent	Technology used	USE (WT/WWT)	Recommendations
		desalination by membrane distillation		water from simulated seawater by integrating a membrane distillation module with a solar still.					used for both brine heating and potable water production. Fluxes of water from both the solar still and the membrane module reached their maximum values 2–3h after the solar irradiation peak.
Mohsen, M.S. , Jamal O., Jaber, J.O., Afonso, M.D.	2003	Desalination of brackish water by nanofiltration and reverse osmosis	Desalination 157 (2003) 167	This study contributes for the development of efficient technologies for the production of affordable potable water in Mediterranean countries where the threat of water shortages is a serious problem, especially in summer.	Jordan	NA	RO &NF	WT/Brackish water	The results show that both processes are efficient, as they highly reduce the organic and inorganic contents present in the raw waters.
Al-Omari, A., Fayyad, M.	2003	Treatment of domestic wastewater by subsurface flow constructed wetlands in Jordan	Desalination 155 (2003) 27–39	The use of subsurface flow constructed wetlands for treating domestic wastewater in Jordan is described.	Jordan	European Union	Subsurface flow constructed wetlands	WWT/domestic	The use of subsurface flow constructed wetlands has reduced BOD. Total nitrogen reduction was observed, nitrate-nitrogen reduction was also observed, and ammonia-nitrogen was reduced by the beds. TSS was reduced by the beds. Total coliform as well as faecal coliform were reduced by one to three logs in all the beds.
Afonso, M.D., Jaber, J.O., Mohsen, M.S.	2004	Brackish groundwater treatment by reverse osmosis in Jordan	Desalination. Vol. 164, no. 2, pp. 157–171. Apr. 2004	Desalination of Red Sea water by reverse osmosis (RO) and/or brackish groundwater desalination by nanofiltration or RO might be technically and economically viable to cope with water scarcity and overcome the water deficit in Jordan.	Jordan	NA	UF-RO	WT (Sea water, brackish groundwater)	The results showed that RO is actually efficient since it highly reduced the content of organic and inorganic matters present in raw waters at a relatively affordable price.

(*continued*)

Authors	Year	Title	Journal name, volume (issue)	Abstract	Country	Funding Agent	Technology used	USE (WT/WWT)	Recommendations
Mohsen, M.S., Jaber, J.O.	2004	Treatment and reuse of industrial effluents: Case study of a thermal power plant	Desalination 167 (2004) 75–86	The potential of industrial wastewater reuse in Jordan's A1-Hussein thermal power station (HTPS) was investigated.	Jordan	NA	RO	WWT/Industrial	It is recommended increasing the capacity of the RO plant, thereby reducing the use of the IE unit. Thus, the consumption of hydrochloric acid and sodium hydroxide is reduced significantly.
Schiffier, M.	2004	Perspectives and challenges for desalination in the 21st century	Desalination 165 (2004) 1–9	The World Bank has undertaken a regional study that aims at improving the understanding of desalination within the Bank and among some of its clients in the Middle East, North Africa and Central Asia.	Jordan	Bank- Netherlands Water Partnership	RO	WT/Sea water	The legal basis and institutional capacity for environmental assessments in general is weak and there is no or very little experience with environmental assessments of individual desalination projects.
Mohsen, M.S.	2004	Treatment and reuse of industrial effluents: Case study of a thermal power plant	Desalination 167 (2004) 75–86	In this paper the potential of industrial wastewater reuse in Jordan's Al-Hussein thermal power station (HTPS) was investigated.	Jordan	NA	RO	WWT/Industrial	It is recommended increasing the capacity of the RO plant, thereby reducing the use of the IE unit. Thus, the consumption of hydrochloric acid and sodium hydroxide is reduced significantly.
Kimura, K., Hane, Y., Watanabe, Y., Amy, G., Ohkuma, N.	2004	Irreversible membrane fouling during ultrafiltration of surface water	Water Research 38 (2004) 3431–3441	In this study, a membrane that had experienced long term filtration of a surface water used as drinking water source was investigated in terms of the recovery in water permeability by chemical cleaning and the composition of membrane foulant desorbed from the membrane. Based on the results obtained in	Jordan	This study was supported by grant-in-aid for Developmental Scientific Research from the Ministry of Education, Science and Culture of Japan.	UF	WT/Surface water	It was demonstrated that the irreversible fouling was caused mainly by organic matter. Metals such as iron and manganese presumably caused the irreversible fouling to some extent. Important finding in this research was that the organic matter contained in the raw after did not uniformly adsorb on the membrane and cause the irreversible fouling.

(*continued*)

Authors	Year	Title	Journal name, volume (issue)	Abstract	Country	Funding Agent	Technology used	USE (WT/WWT)	Recommendations
				this study, it will be discussed which constituents in the raw water caused the irreversible membrane fouling.					
Al-Jayyousi, O.	2004	Greywater reuse: knowledge management for sustainability	Desalination 167 (2004) 27–37	This paper aims to highlight the value of knowledge management in greywater reuse.	Jordan	NA	NA	WWT/Greywater	Leveraging the tacit knowledge base of an individual and making use of socialization to transfer it throughout the organization. Amplifying knowledge creation across different levels of the organization. Enhancing the enabling conditions and Continuing to create new knowledge constantly.
Al-Hayek, I., Badran, O.O.	2004	The effect of using different designs of solar stills on water distillation	Desalination 169 (2004) 121–127	The objective of this paper is to improve the quality and increase the quantity of drinking water for people living in remote desert areas in Jordan.	Jordan	NA	Desalination	WT/solar water	It can be concluded that the use of solar water distillation promises to enhance the quality of life and to improve health standards in arid areas in Jordan.
Abu Qdais, H., Moussa, H.	2004	Removal of heavy metals from wastewater by membrane processes: a comparative study.	Desalination 164 (2004) 105–110	The purpose of this study is to evaluate the performance of the RO and NF processes in removing heavy metals from industrial wastewater.	Jordan	NA	RO&NF	WWT/Industrial	Membrane process should be explored further for the possibility of heavy metals recovery.
Hilal, N., Al-Zoubi, H., Mohammad, A.W., Darwish, N.A.	2005	Nanofiltration of highly concentrated salt solutions up to seawater salinity	Desalination 184 (2005) 315–326	The main parameters studied in this paper are feed pressure and salt concentration. The experimental data of this work was correlated and analysed using the Spigler-Kedem model.	Jordan	MEDRC	NF	WT/Sea water	NF membranes have the potential to be used as a pre-treatment alternative in desalination processes.

(continued)

Authors	Year	Title	Journal name, volume (issue)	Abstract	Country	Funding Agent	Technology used	USE (WT/WWT)	Recommendations
Abdullah, S., Abu-Hilala, M., Mohsen, M.S.	2005	Performance of a photovoltaic powered reverse osmosis system under local climatic conditions	Desalination 183 (2005) 95–104	An experimental study was conducted to investigate the potential of the development of water desalination using photovoltaic powered system in Jordan.	Jordan	NA	RO	WT/(seawater and brackish water)	The photovoltaic powered RO system has a favourable application. More experimental work needs to be carried out to study the continuous performance of the system, and more investigations should be directed to the membrane fouling and recovery ratio of the system.
Ammary, B.Y.	2005	Treatment of olive mill wastewater using an anaerobic sequencing batch reactor	Desalination 177 (2005) 157–165	Wastewater from olive mills in Jordan has been characterized and treated using a lab scale anaerobic sequencing batch reactor (ASBR).	Jordan	NA	Lab scale anaerobic sequencing batch reactor (ASBR)	Olive mill wastewater	Dilution of the OMW was necessary for the anaerobic treatment of such wastewater. This was achieved by adding raw OMW to the reactor containing treated OMW from the previous batch. This, in addition to no nutrient requirement, substantially reduces the cost of treatment.
Al-Rawajfeh, A.E.	2007	Modelling of alkaline scale formation in falling film horizontal-tube multiple-effect distillers	Desalination 205 (2007) 124–139	In the present study, a model was developed, coupling mass transfer with chemical reaction, to calculate alkaline scale formation from simultaneous desorption–deposition of CO_2–$CaCO_3$ from evaporation of water falling film in horizontal-tube ME distillers.	Jordan	Prof. Ibraheem Jibreel, Tafilah Technical University.	Multiple-effect (ME) distillers	WT	The CO_2 desorption rates as well as $CaCO_3$ deposition rates, scale thickness and fouling resistance Rf were investigated at different top brine temperatures (TBT), concentration factor (CF), pH and salinity.
Al-Qodah, Z., Shawaqfeh, A.T., Lafia, W.K.	2007	Adsorption of pesticides from aqueous solutions using oil shale ash	Desalination 208 (2007) 294–305	Adsorption characteristics of the pesticides Deltamethrin and Lambda-Cyhalothrin	Jordan	VAPCO (Amman, Jordan), for Prince Faisal Dead Sea Studies Center (Mutah University, Karak,	Oil shale ash (OSA)	Pesticides	The results of the desorption experiments indicate that acetate buffer can remove the pesticides from the OSA particles.

(continued)

Authors	Year	Title	Journal name, volume (issue)	Abstract	Country	Funding Agent	Technology used	USE (WT/WWT)	Recommendations
				were studied in aqueous solutions using oil shale ash (OSA) as an adsorbent.		Jordan) and to the Authority of Natural resources (Amman, Jordan).			This gives a method for both recovering the pesticides and regenerating the adsorbent.
Darwish, M.R., Shararaa, M., Sidahmeda, M., Haidar, M.	2006	The impact of a storage facility on optimality conditions of wastewater reuse in land application: A case study in Lebanon	Resources, Conservation and Recycling, In Press, Corrected Proof, Available online 3 November 2006	Investigate the economic feasibility of land application in the presence of a reservoir with different capacities and to compare the outcome to the application without a storage reservoir.	Lebanon	NA	Land application	WWT	
Kershman, S.A.	2001	25 years of experience in operating thermal desalination plants	Desalination 136 (2001) 141–145	This paper describes a complete survey of all installed thermal desalination plants and also describes the problems encountered in the past and the adapted solutions for some of these problems.	Libya	GECOL	MSF&MED	WT/Sea water	Desalination in Libya has good prospects and plays an important role in securing water for urban use. The country invests in this field, and many efforts are made to reduce the specific water production cost.
Bindra, S.P., Abosh, W.	2001	Recent developments in water desalination	Desalination 136 (2001) 49–56	The paper presents a global overview of the desalination of seawater.	Libya	Eng. All Sad Ballgames	Desalination	WT/SW	There is a vital need to develop new water policies outlining measures to bridge the gap between current policies and those that lead to the sustainable use of the world's water resources.
Wheida, E., Verhoeven, R.	2004	Desalination as a water supply technique in Libya	Desalination 165 (2004) 89–97	This paper summarises the techniques, their installation capacities and the associated problems in running and maintaining desalination industrial plants, as they will become a vital source of water supply in Libya.	Libya	NA	Desalination	W&WWT	There is need for a national water strategy to bridge the gap between the actual water policy and a strategy that leads to sustainable use of the country's water resources. The installation of suitable capacities to meet the water demands will certainly play an important role in sustaining the economic development of the country.

(*continued*)

Authors	Year	Title	Journal name, volume (issue)	Abstract	Country	Funding Agent	Technology used	USE (WT/WWT)	Recommendations
Abouswa, K., Elshawesh, F., Elragei, O., Elhood, A.	2007	Corrosion investigation of Cu–Ni tube desalination plant	Desalination 205 (2007) 140–146	Copper-based alloy tubes (i.e. Cu–Ni) are extensively used in desalination plants. This is in order to complete the heating and evaporating process for the seawater and to obtain distillate water that can be used in the steam generation plant and as drinking water.	Libya	NA	Desalination	WT/SW	Identify the source of seawater salts found on Cu–Ni tubes as a result of internal corrosion leakage or as a result of demister/vacuum problem. The presence of chloride ions was confirmed by EDAX analysis. Check the performance of the vacuum system of the desalination unit(s). This is an important step and need to be considered. Presence of gases shall assist in initiation and acceleration in the corrosion process.
Raissouni, I., Marraha, M., Azmani, A.	2007	Effect of some parameters on the improvement of the bipolar membrane electrodialysis process	Desalination 208 (2007) 62–72	A study of electrodialysis with bipolar membranes of $NaNO_3$ solutions to produce nitric acid and soda was performed with the object of stressing the role of the auxiliary ion-exchange membranes used to equip the bipolar membrane electrodialysis cell.	Morocco	The authors are grateful to Professors H.D. Hurwitz and C. Buess from the Université Libre of Brussels (ULB, Belgium)	ED	HNO_3 and NaOH	The generation of HNO_3 and NaOH by electrodialysis starting from a $NaNO_3$–NaCl mixed salt solution proves that the recovered HNO_3 contamination increases only slightly with the increasing concentration of NaCl.
Assaf, S.A.	2001	Existing and the future planned desalination facilities in the Gaza Strip of Palestine and their socio-economic and environmental impact	Desalination 138 (2001) 17–28	One of the major options for the remedy of water shortages is the utilization of desalination technology for brackish and seawater in Gaza Strip. On the other hand, the brine disposal from the private desalination plants and the small	Palestine	NA	RO	WT/Sea water	The badly needed research on solving the anticipated problems in operation, maintenance and efficiency of seawater desalination should be started now without delay. The RO method which is the most appropriate and most economic for the precarious situation in the Gaza Strip.

(*continued*)

Authors	Year	Title	Journal name, volume (issue)	Abstract	Country	Funding Agent	Technology used	USE (WT/WWT)	Recommendations
				private RO water vendors are a menacing and uncontrolled environmental problem in the Gaza Strip.					
Abu-Jabal, M.S., Karniyab, I., Namsakib, Y.	2001	Proving test for a solar-powered desalination system in Gaza-Palestine	Desalination 137 (2001) 1–6	This paper describes the unit, discusses the method used for distillation and results of the tests during the testing period. It also shows the feasibility of using the system in our region.	Palestine	The Ebara Corporation (Tokyo) and the Water Research Center in Al Azhar University	Solar radiation	WT/Sea water	The system can be used in isolated areas, (zero-brine-discharge) system. Operation of the system is very easy. Productivity for this system is higher than other solar desalination processes. High corrosion-resistant material. Maintenance for the system is not significant; a number of units can be used in parallel. The water quality produced is very high.
Abu Zahra, B.A.A.	2001	Water crisis in Palestine	Desalination 136 (2001) 93–99	Palestine is experiencing a severe water crisis caused mainly by the lack of control over the Palestinian water resources.	Palestine	NA	Desalination	W&WWT	The overall objective of any integrated water resources management plan is to satisfy the needs for sustainable development.
El Sheikh, R., Ahmed, M., Hamdan, S.	2003	Strategy of water desalination in the Gaza Strip	Desalination 156 (2003) 39–42	Reverse osmosis desalination is strongly recommended and considered as a strategic alternative in order to overcome the water deficit and meet the future needs of desalination for the Strip.	Palestine	NA	RO	WT/Sea water	It is recommended to have large-scale SWRO plants in order to minimize the operation costs which are relatively higher than in the small ones.
Jaber, I.S., Ahmed, M.R.	2004	Technical and economic evaluation of brackish groundwater	Desalination 165 (2004) 209–213	Due to technical and economic considerations, desalination of	Palestine	NA	RO	WT(Brackish & Sea water)	The total production cost is still high. The local operators of RO plants have more knowledge to

(*continued*)

Authors	Year	Title	Journal name, volume (issue)	Abstract	Country	Funding Agent	Technology used	USE (WT/WWT)	Recommendations
		desalination by reverse osmosis (RO) process		brackish water and/or seawater by RO technology was to chosen to solve water problems in terms of quantity and quality.					run their plants than those in BWRO plants and the water market is limited. The private sector should be encouraged in water desalination.
Hilal, N., Al-Zoubi, H., Darwish, N.A., Mohammad, A.W., Abu Arabi M.	2004	A comprehensive review of nanofiltration membranes: Treatment, pretreatment, modelling, and atomic force microscopy	Desalination 170 (2004) 281–308	This paper reviews several aspects of NF membranes such as the applications in treatment of ground, surface, and wastewater; as a pretreatment for desalination, fouling, and the separation mechanism; and modelling of NF membranes.	Palestine	MEDRC	NF	WT7WWT	AFM has been mentioned in this review, which can significantly characterize the surface of NF membranes. Surface morphology, pore size distribution and particle adhesion of NF membrane were able to be quantified by AFM.
Al-Jamal, K.H., Shoblak, M.E.	2004	Costing a Major Desalination Plant for Gaza on the Mediterranean Region	International conference on desalination costing, Dec, 2004, Cyprus	This paper summarizes appropriate configurations for large scale reverse osmosis (RO) seawater desalination plant on the Mediterranean.	Palestine	NA	RO	WT/Sea water	In order to reach a well informed decision, desalination is discussed in the context of integrated water resource management for Gaza region.
Khamis, M., Manassra, A., Dakiky, M.	2005	Wastewater treatment by advanced membrane Technology	International conference: water values and rights	In order to obtain effluent that can be used for non-restricted irrigation and sustainable application to plant and soil a tertiary treatment pilot plant was installed. The technology used based on ultra-filtration and reverse osmosis.	Palestine	PCG & Belgian Government	UF-RO	WWT/ Municipalities	UF permeate is free from all pathogens. The RO permeate is free of major ions including chlorides. More work is required to verify the effect of changing operational parameters on water quality.
AL-Agha, M.R., Mortaja, R.S.	2005	Desalination in the Gaza Strip: drinking water supply	Desalination 173 (2005) 157–171	Strict policies and management actions are to be taken in order to alleviate the	Palestine	NA	RO	WT (brackish & Sea water)	Brine management and maintaining of brackish and seawater desalination plants are very important.

(*continued*)

Authors	Year	Title	Journal name, volume (issue)	Abstract	Country	Funding Agent	Technology used	USE (WT/WWT)	Recommendations
		and environmental impact		health and environmental impact of an important and uncontrolled new system.					
Baalousha, H.	2006	Desalination status in the Gaza Strip and its environmental impact	Desalination, Volume 196, Issues 1–3, 5 September 2006, Pages 1–12	Desalination seems to be the only viable alternative for water resources.	Palestine	NA	RO	WT/Sea water	The quality of desalinated water must be monitored and the impact of desalination on the environment investigated, and possible measures must be taken to reduce this impact.
Sbeih, M.Y.	1996	Recycling of treated water in Palestine: Urgency, obstacles and experience to date	Desalination 106 (1996) 165–178	To provide piped water to rural areas of Palestine, low technology treatment plants and sewers should be installed at minimum cost and for ease of operation.	Palestine	NA	Desalination	WT&WWT	Public education and awareness should be carried out to discuss the idea of reuse to the public to clear the idea.
Assaf, S.A.	2001	Existing and the future planned desalination facilities in the Gaza Strip of Palestine and their socio-economic and environmental impact	Desalination 138 (2001) 17–28	One of the major options for the remedy of water shortages is the utilization of desalination technology for brackish and seawater in Gaza Strip. On the other hand, the brine disposal from the private desalination plants and the small private RO water vendors are a menacing and uncontrolled environmental problem in the Gaza Strip.	Palestine	NA	RO	WT/Sea water	The badly needed research on solving the anticipated problems in operation, maintenance and efficiency of seawater desalination should be started now without delay. The RO method which is the most appropriate and most economic for the precarious situation in the Gaza Strip.
Abu Zahra, B.A.A.	2001	Water crisis in Palestine	Desalination 136 (2001) 93–99	Palestine is experiencing a severe water crisis caused mainly by the lack of	Palestine	NA	Desalination	W&WWT	The overall objective of any integrated water resources management plan is to satisfy the needs

(*continued*)

Authors	Year	Title	Journal name, volume (issue)	Abstract	Country	Funding Agent	Technology used	USE (WT/WWT)	Recommendations
				control over the Palestinian water resources.					for sustainable development.
El Sheikh, R., Ahmed, M., Hamdan, S.	2003	Strategy of water desalination in the Gaza Strip	Desalination 156 (2003) 39–42	Reverse osmosis desalination is strongly recommended and considered as a strategic alternative in order to overcome the water deficit and meet the future needs of desalination for the Strip.	Palestine	NA	RO	WT/Sea water	It is recommended to have large-scale SWRO plants in order to minimize the operation costs which are relatively higher than in the small ones.
Teixeira, M.R., Rosa, M.J.	2006	Neurotoxin and hepatotoxic cyanotoxins removal by nanofiltration	Water Research 40 (2006) 2837–2846	This study investigates the influence of chemical feed characteristics on nanofiltration performance for cyanotoxins removal, namely the neurotoxic anatoxin-a (alkaloid of 166g/mol, positively charged) and the hepatotoxic microcystins (cyclic peptides of 1000g/mol, negatively charged).	Portugal	Prodep Program	NF&RO	WT/Natural water	Microcystins in the NF permeate were always below the quantification limit, hence far below the WHO drinking water guideline value of 1mg/L for MC-LR.
Borja, R. Banks, C.J., Garrido, A.	1994	Kinetics of Black-Olive Wastewater Treatment by the Activated-Sludge System	Process Biochemistry 29 (1994) 587–593	The objective of this work was to investigate the feasibility of using a completely mixed activated-sludge (AS) system to treat black-olive wastewater.	Spain	Departamento de Postgrado y Especializacion of the Spanish CSIC	Activated sludge	WWT/Black-olive	Black-olive wastewater is easily degraded by the completely mixed activated sludge treatment system.
Euglena, A.M., Nunez, A., Nishid, M.	1999	Experimental test of TOYOBO membranes for seawater desalination at Las Palmas, Spain	Desalination 125 (1999) 55–64	This study's focus is to confirm the performance of the Hollosep membrane when it is operated under continuous operation, high recovery ratio and feeding open sea	Spain	CEDEX & EMALSA	RO	WT/SW	Hollosep is a high chlorine resistance membrane which can present high quality and stable volume ratio of permeate water for a long time operation for RO desalination plants having open sea intake water. Hollosep can

(*continued*)

Authors	Year	Title	Journal name, volume (issue)	Abstract	Country	Funding Agent	Technology used	USE (WT/WWT)	Recommendations
				intake water at Las Palmas, Gran Canaria Is., Spain.					perform long-term operation without chemical treatments due to CTA membrane material which has high chlorine tolerance. The Hollosep brings seawater desalination plant operation easily using CCl or ICl.
Teulera, A., Glucina, K., Laîné, J.M.	1999	Assessment of UF pretreatment prior RO membranes for seawater desalination	Desalination 125 (1999) 89–96	The aim of this study was to assess an integrated membrane system, i.e. combined membrane processes to treat surface seawater. The main objectives were to: • Compare spiral wound thin film composite (SW) membrane to the hollow fine fiber (HFF) membrane for the production of drinking water from beach-well seawater; • Evaluate the ultrafiltration performances on high TDS water; • Conduct a long-term investigation to determine the efficiency of an integrated membrane system UF/RO for surface seawater	Spain	Suez-Lyonnaise des Eaux, CPERI of the University of Thessaloniki, AGBAR of Barcelona, Mekorot Water Company and the Water Research Institute (WRI) of the Technion – Israel Institute of Technology. Permission to publish this paper is gratefully acknowledged.	RO & UF	WT/SW	Further trials are on-going (flux at 140L/h.m^2) in order better define the limits of the UF membrane. Following this test, UF will be tested on surface intake and the UF water will feed the RO pilot. The RO pilot tests will be conducted side by side to evaluate the impact of the pretreatment on the RO hydraulic performances.

(continued)

Authors	Year	Title	Journal name, volume (issue)	Abstract	Country	Funding Agent	Technology used	USE (WT/WWT)	Recommendations
				desalination in comparison with beachwell/RO treatment process.					
Vera, L., Delgado, S., Elmaleh, S.	2000	Dimensionless numbers for the steady-state flux of cross-flow microfiltration and ultrafiltration with gas sparging	Chemical Engineering Science 55 (2000) 3419–3428	The aim of the present work is to apply tangential diphasic gas liquid flow under conditions, which were not already studied, and extend the dimensionless approach to sparged cross-flow filtration.	Spain	NA	UF&MF	WWT/Industrial	The dimensionless quantities introduced in conventional cross-flow filtration can be extended to gassparged filtration.
Rodríguez, J.J., Jiménez, V., Trujillo, O., Veza, J.M.	2002	Reuse of reverse osmosis membranes in advanced wastewater treatment	Desalination 150 (2002) 219–225	This paper describes the results of preliminary tests of membrane peeling, made on a small test unit, and followed by tests in a full pilot plant to be built.	Spain	Empresa Mixta de Aguas de Las Palmas SA (EMALSA)	RO	WWT/domestic	The next steps will be aimed at testing the actual filtration capability of the peeled membranes as related to suspended matter in actual wastewater. For that purpose a new plant for tests with treated wastewater under real plant conditions will soon be built. It is also expected to specify more precisely the optimal dose of potassium permanganate to be used.
Bodalo-Santoyo, A., Gomez-Carrasco, J.L., Gomez-Gomez, E., Maximo-Martin, F., Hidalgo-Montesinos, A.M.	2003	Application of reverse osmosis to reduce pollutants present in industrial wastewater	Desalination 155 (2003) 101–108	We describe a comparative study of four different polyamide membranes that were tested for their ability to reduce the concentration of pollutants in a synthetic effluent stream containing acrylnitrile and three inorganic species (sulphate, ammonium and cyanide).	Spain	CICYT, Madrid (Spain)	RO	WWT/Industrial	Cyanide and ammonium cannot be acceptably eliminated in a single step operation when they are simultaneously present in industrial wastewater. The results point to the need to carry out the different separation steps at different pH values.

(*continued*)

Authors	Year	Title	Journal name, volume (issue)	Abstract	Country	Funding Agent	Technology used	USE (WT/WWT)	Recommendations
Bodalo-Santoyo, A., Gomez-Carrasco, J.L., Gomez-Gomez, E., Maximo-Martin, M.F., Hidalgo-Montesinos, A.M.	2004	Spiral-wound membrane reverse osmosis and the treatment of industrial effluents	Desalination 160 (2004) 151–158	This work forms part of a wider research project aimed at recovering products and reducing the concentration of pollutants in wastewater by using membrane process.	Spain	CICYT, Madrid (Spain).	RO	WWT/Industrial	The sulphate ion shows high rejection values, NH: and CN- elimination shows a strong dependence on feed pH. Not being an ionic species, acrylnitrile shows low rejection percentages in all the steps.
Blanco, J., Alarcón, D.	2004	Innovation ideas to reduce current cost of solar seawater desalination based on med technology	julian. blanco@psa.es	The project aiming to the further enhancement of economic efficiency and environmental protection.	Spain	NA	Desalination (MSF)	WT/Sea water	After the analysis of the results obtained for the complete reference year, the external mixing configuration with heat pump start-up and shutdown was shown a priori to be the most satisfactory configuration with regard to temperature and flow rate obtained.
Bodalo, A., Gomez, J.L., Gomez, E., Hidalgo, A.M., Aleman, A.	2005	Viability study of different reverse osmosis membranes for application in the tertiary treatment of wastes from the tanning industry	Desalination 180 (2005) 277–284	This paper describe a comparative study of six different reverse osmosis membranes, which were tested for their ability to reduce the salt content in the tertiary treatment after the elimination of chromium salts and organic matter of an effluent from a pilot plant for treating industrial wastewater from the tanning industry to reach the legal levels established for their safe disposal.	Spain	Aquagest Levante S.A., Murcia (Spain)	RO	WWT/Industrial	DESAL-3B membrane provided the lowest conductivity values in the permeate stream, and so this membrane was selected for use in the reverse osmosis pilot step following the bioreactor in the processing of the industrial tanning wastewater.
Sancho, M., Arnal, J.M., Pineda, A. Catalfi, R., Lora, J.	2005	Application of membrane technology for the treatment of effluent from a zirconium	Desalination 178 (2005) 361–367	This work studied the application of membrane technology (basically UF) for the separation of polyelectrolytes from	Spain	NA	UF	WWT/domestic	The spill water to be treated presents a high heterogeneity. It is necessary to establish a system of homogenization of the

(*continued*)

Authors	Year	Title	Journal name, volume (issue)	Abstract	Country	Funding Agent	Technology used	USE (WT/WWT)	Recommendations
		silicate production process		the spill water of this company, with two important purposes: (a) the reusability of the polyelectrolytes in the process of decanting and (b) to reuse the water in the milling process to diminish the amount of spill water.					characteristics of the spill water in order to define and control in an effective way conditions of the later treatment. Filtration is a suitable method to eliminate suspended solids; in the case of zirconium the tests with acetate cellulose flat filters have shown very successful results. Due to its capacity of retention, the most suitable pore size for the filtration of the spill water is 0.45m. Filtration is an optimal pretreatment to the application of membrane technology (ultrafiltration) to spill water since it removes all the suspended solids that could block membranes and decrease their yield.
Santafe-Moros, A., Goza lvez-Zafrilla, J.M., Lora-Garcia, J.	2005	Performance of commercial nanofiltration membranes in the removal of nitrate ions	Desalination 185 (2005) 281–287	The aim of the present work was to study the efficiency of the nitrate removal by three commercial nanofiltration membranes, NF90, NF270 (Dow-FilmTec) and ESNA1-LF (Hydranautics).	Spain	Dow-FilmTec and Hydranautics	NF	WT/Ground water	This confirms the convenience of experimental tests on pilot plant with the groundwater to be treated in order to select the most best-suited membrane.
Oyanedel, V., Campos, J.L., Garrido, J.M., Lazarova, V., Mendez, R.	2005	Development of a membrane-assisted hybrid bioreactor for ammonia and COD removal in wastewaters	Chem Technol Biotechnol 80:206–215 (2005)	A new membrane-assisted hybrid bioreactor was developed to remove ammonia and organic matter	Spain	Ministry of Science and Technology, Spain	Hybrid bioreactors	WWT/Municipal	The use of a membrane had the following advantages over the previous design of the system with a settler: best quality of the effluent in terms of ammonia concentration; best control of the suspended biomass concentration in the reactor.

(*continued*)

Authors	Year	Title	Journal name, volume (issue)	Abstract	Country	Funding Agent	Technology used	USE (WT/WWT)	Recommendations
López-Ramírez, J.A., Coello Oviedo, M.D., Quiroga Alonso, J.M.	2006	Comparative studies of reverse osmosis membranes for wastewater reclamation	Desalination 191 (2006) 137–147	The goal of this paper is to show part of the studies conducted using an experimental pilot plant (output: $100 m^3/d$) in Chiclana de la Frontera, Province of Cadiz, southern Spain. The purpose pilot-plant study was to define the optimum conditions for physicochemical pretreatment of secondary effluents for successful reverse osmosis operation for groundwater recharge.	Spain	PRIDESA	RO	WWT/domestic	Membrane surface morphology plays a paramount role in colloidal deposition. PA membranes have a higher permeability than CA membranes. High recoveries lead to a fast and usually irreversible fouling in membranes. An adequate pre-treatment is of utmost importance for correct membrane performance.
Gómez, M., de la Rua, A., Garralón, G., Plaza, F., Hontoria, E., Gómez, M.A.	2006	Urban wastewater disinfection by filtration technologies	Desalination 190 (2006) 16–28	The aim of the present study was to assess the microbiological quality (based on the parameters faecal coliform, E. coli, coliphages and nematode eggs) of different effluents from two macrofiltration processes (pressure sand filter and disc filter) used as pretreatment and membrane technologies (microfiltration and ultrafiltration), in order to determine the possibility of using filtration as an alternative to wastewater disinfection.	Spain	CADAGUA	MF&UF	WWT/Municipal	Both macrofiltration systems used in the experiment (disc filter and pressure sand filter) produced effluent without pathogenic nematode eggs. Effluent from both the disc filter system and the sand pressure system is of variable quality and depends on the quality of influent. False positives were detected in faecal coliform results. The best results were achieved by the ultrafiltration module.

(*continued*)

Authors	Year	Title	Journal name, volume (issue)	Abstract	Country	Funding Agent	Technology used	USE (WT/WWT)	Recommendations
Urtiaga, A., Abellán, M.J., Irabien, A., Ortiz, I.	2006	Use of membrane contactors as an efficient alternative to reduce effluent ecotoxicity	Desalination 191 (2006) 79–85	This works reports on the separation and recovery of copper cations from water streams from the wet peroxide oxidation process (WPO) using emulsion pertraction technology (EP) in hollow-fiber contactors as separation technology.	Spain	Ministry of Education and Science (Spain)	Hollow-fiber contactors	WWT/Industrial	This works reports the application of a HF membrane contactor to the separation and recovery of Cu^{2+} from aqueous solutions using the EP process.
Arnal, J.M., García-Fayos, B., Sancho, M., Lora, J., Verdú, G.	2006	Water potabilization in developing countries: membrane technology and natural coagulants	Desalination 200 (2006) 325–326	Surface water contains very high pathogenic microbial contaminants, mainly bacteria and virus, due to animal and organic matter decomposition. Intake of the raw water is a high risk for health that affects mainly children and old people.	Spain	LAINSA, Centro Cooperación al Desarrollo from UPV, Hydrology and Natural Medium Institute from UPV, Naquera town (Valencia), Vicerrectorado de Cooperación y Desarrollo from UPV, Spanish non-Gubernamental organization (ACOEC) and Ecuadorian non-Gubernamental organization (SER Foundation)	UF	WT/Surface water	Natural coagulants from oil seeds show promising results in coagulation and antimicrobial test. Agro-based materials can be an exploited option towards achieving sustainable drinkable water.
Rivero, M.J., Parsons, S.A., Jeffrey, P., Pidou, M., Jefferson, B.	2006	Membrane chemical reactor (MCR) combining photocatalysis and microfiltration for grey water treatment	Water Science & Technology Vol. 53 No. 3 pp. 173–180	This paper looks at reusing grey water; the preference is treatment processes based on biological systems to remove the dissolved organic content	Spain	Fundación Marcelino Botín	MCR&MF	WWT/greywater	Future studies will try to achieve working with higher permeate fluxes as it has not been possible yet to reach a critical flux. Organic matter removal and grey water turbidity will be monitored to ensure that the desired reduction is still achieved.
Martínez, L., Rodríguez-Maroto, J.M.	2007	Effects of membrane and module design improvements on flux	Desalination 205 (2007) 97–103	The aim of this work is to show the effects of improving membrane and module	Spain	NA	DCMD	WT	If high fluxes are targeted, both membrane and module characteristics must be adequate. If this is

(continued)

Authors	Year	Title	Journal name, volume (issue)	Abstract	Country	Funding Agent	Technology used	USE (WT/WWT)	Recommendations
		in direct contact membrane distillation		characteristics on flux when different feeds are processed.					not so, the good characteristics of only one of them (module or membrane) will not produce the desired flux because its good characteristics can be overshadowed by the inadequate behaviour of the other one.
Colella, F., Uche, J., Dejoc, N.	2007	Desalted water provided by a poligeneration scheme for the tourist sector	Desalination 205 (2007) 279–297	The aim of the paper is to present a poligeneration scheme that provides desalted water in a reliable and economic manner for the tourist sector, so contributing to show that the concept of 'distributed generation' for water is also possible, as electricity is.	Spain	NA	Desalination	WT	The case study applied in this paper demonstrates that the combined production of many products is advantageous both in terms of primary energy saving (less CO^2 emissions) and economical profitability.
Gómez, M., Plaza, F., Garralón, G., Pérez, J., Gómez, M.A.	2007	A comparative study of tertiary wastewater treatment by physico-chemical-UV process and macrofiltration–ultrafiltration technologies	Desalination 202 (2007) 369–376	The objective of the present study is to compare physico-chemical UV treatment and membrane technologies with regard to the final quality of treated effluent and the associated economic costs, with a view to establishing the applicability of each system.	Spain	CADAGUA., Basque Government's INTEK programme	MF & UF	WWT/domestic	At present the membrane system remains less competitive than physico-chemical-UV treatment, owing particularly to the high cost of the installation and membrane replacement.
Mandil, M.A., Bushnak, A.A.	2002	Future needs for desalination in South Mediterranean countries	Desalination 152 (2003) 15–18	This is an attempt to forecast the role desalination is expected to play	Syria	Dar Al-Taqniya Institute, UNESCO Regional Training and Water Studies Center,	Desalination	WT & WWT (seawater, brackish and wastewater)	The need for environmental sustainability, capacity building in desalination

(*continued*)

Authors	Year	Title	Journal name, volume (issue)	Abstract	Country	Funding Agent	Technology used	USE (WT/WWT)	Recommendations
				during the first quarter of the 21st century in the south Mediterranean countries.		Cairo, Mohammed Bin Rashid Technology Park, Dubai, Makkah Technology Park			technology and financial sustainability.
Deane, J.	2002	Private sector participation in desalination in the Mediterranean Middle East (MIME) – past, present and future.	Desalination 152 (2002) 57–66	This paper considers the scope for Private Sector Participation (PSP) in desalination projects in the Mediterranean Middle East (MME) countries.	Syria	NA	Desalination	WT & WWT (seawater, brackish and wastewater)	In general the demand for desalination capacity is driven by the lack of economically viable alternatives. The MME countries identified as requiring additional desalination capacity in the foreseeable future.
BenJemaa, F., Houcine, I., Chahbani, M.H.	1998	Desalination in Tunisia: Past experience and future prospects	Desalination 116 (1998) 123–134	This paper covers a number of ambitious initiatives, which tackled the water shortage problem in Tunisia through the adoption of original programs and solutions adapted to the local environment. Presented are also the prospects of water desalination in Tunisia, which is currently pursued as a major component of an overall strategy to meet future water demand and enhance the quality of drinking water in various parts of the country.	Tunisia	NA	RO	WT/SW	Given the presented statistics on the needs and availability of water and energy, coherent regional strategies have to be elaborated to face future water shortages in the Mediterranean region with desalination being a major component. It is high time to seriously consider the danger of an impending water shortage with its dramatic social and economic consequences.
Ben Bacha, H., Maalej, A.Y., Ben Dhia, H., Ulber, I., Uchtmann, H., Engelhardt, M., Krelle, J.	1999	Perspectives of solar-powered desalination with the “SMCEC” technique	Desalination 122 (1999) 177–183	In this article, the survey, description and perspectives of a new solar-based desalination technique – solar multiple condensation evaporation cycle (SMCEC) – are presented.	Tunisia	NA	SMCEC	WT/BW	Medium and small desalination units are a viable solution to high energy savings due to the fact that solar energy is free and that the solar distillation systems do not need complex technology.

(*continued*)

Authors	Year	Title	Journal name, volume (issue)	Abstract	Country	Funding Agent	Technology used	USE (WT/WWT)	Recommendations
Sayadi, S., Allouche, N., Jaoua, M., Aloui, F.	2000	Detrimental effects of high molecular-mass polyphenols on olive mill wastewater biotreatment	Process Biochemistry 35 (2000) 725–735	To investigate whether low, medium or high molecular-mass polyphenolics were the most problematic compounds for olive mill wastewater (OMW) biotreatments.	Tunisia	EEC&SERST-Programmes Nationaux Mobilisateurs	UF	WWT/(OMW)	Efforts are being focused on reducing concentrations of high molecular mass polyphenols in order to prevent LiP inhibition.
Hamrouni, B., Dhahbi, M.	2001	Analytical aspects of silica in saline waters application to desalination of brackish waters	Desalination 136 (2001) 225–232	Determination of amorphous silica content of natural waters, was carried out by various analytical methods. Amorphous silica solubility, at different temperatures, pH and ionic strength was studied.	Tunisia	NA	Desalination	WT/BW	Solubility of amorphous silica were determined at different temperatures, pH and ionic strength. Heat of silica dissolution and the activity coefficient of dissolved silica were determined.
Dhouib, A., Hamad, N., Hassaïri, L., Sayadi, S.	2003	Degradation of anionic surfactants by Citrobacter braakii	Process Biochemistry 38 (2003) 1245–1250	To select and isolate bacteria or consortia of bacteria able to degrade high concentrations of anionic surfactants rapidly. (ii) to use these strains in a bio-augmentation process for the treatment of a cosmetic industry wastewater and (iii) to connect the bioreactor to a cross-flow micro-filtration membrane unit for the treatment of a detergent and cosmetic industry wastewater and reuse it in situ.	Tunisia	SERST/Programmes Nationaux Mobilisateurs	Membrane bioreactor	WWT/Anionic surfactants	A bacterium with a wide anionic surfactant degradation spectrum was isolated.
Tlili, M.M., Manzola, A.S., Ben Amor, M.	2003	Optimization of the preliminary treatment in a desalination plant by reverse osmosis	Desalination 156 (2003) 69–78	This work exploits the different scaling techniques with the aim of calcium salt prevention.	Tunisia	NA	RO	WT/Sea water	The fouling of RO membranes by the Mg(OH), suspended microflows precipitated by elevated pH due to chemical additions is avoided.

(*continued*)

Authors	Year	Title	Journal name, volume (issue)	Abstract	Country	Funding Agent	Technology used	USE (WT/WWT)	Recommendations
Fergana, E., Roudesli, S., Deratani, A.	2004	Desalination of brackish water from Tunisian Sahel using composite polymethylhydro-siloxane-cellulose acetate membranes	Desalination 162 (2004) 103–109	Composite cellulose acetate (CA) based membranes having a hydrophobic top layer made of polymethyl-hydrosiloxane (PMHS) were prepared with the aim of desalination of brackish waters from the Sahel region of Tunisia having a total dissolved salt content (TDS) of about 4100g/m^3	Tunisia	The French Foreign Office. The French Ministry of Education and The Tunisian Ministry of Education and Scientific Research	Polymethyl-hydro-siloxane-cellulose acetate membranes	WT/BW	It results in the same salt retention – an improvement factor of 50–100% for the permeation rate when using the composite membranes. The ion rejection in decreasing order was $SO_4^{2-} > C^{1-} >$ NOr and $Mg^{2+} > Ca^{2+} > Na^{+}$ for anions and cations, respectively.
Mechichi, T., Sayadi, S.	2005	Evaluating process imbalance of anaerobic digestion of olive mill wastewaters	Process Biochemistry 40 (2005) 139–145	In this study, we have evaluated parameters for process imbalance using an up-flow anaerobic filter reactor containing highly toxic olive mill wastewater.	Tunisia	Hedi Aouissaoui and Nejla Masmoudi (Centre de Biotechnologie de Sfax)&Professor Bharat K.C. Patel (Griffith University, Australia)	Anaerobic filter reactor	WWT/Olive mill	Valerate accumulation was shown to be the most toxic of all VFAs but an accumulation of propionate, butyrate and acetate also inhibited the functioning of the reactor.
Dhouib, A., Hdiji, N., Hassaïri, I., Sayadi, S.	2005	Large scale application of membrane bioreactor technology for the treatment and reuse of an anionic surfactant wastewater	Process Biochemistry 40 (2005) 2715–2720	A powerful Citrobacter braakii strain able to degrade wide range of anionic surfactants.	Tunisia	«Secrétariat d'Etat à la Recherche Scientifique et à la Technologie» and the «COSMART» factory. Dr. Ingo. GEISSEN Sven and Dipl.-Ing. FITZKE Bernd from Technical University of Clausthal.	(UF)Membrane bioreactor	WWT/Industrial	The carbon, nitrogen and surfactant concentration in the effluent met the wastewater reclamation standard in Tunisia. The methods to prevent and reduce membrane fouling should be further investigated in detail.
Hazami, M., Kooli, S., Lazaar, M., Farhat, A., Belghith, A.	2005	Performance of a solar storage collector	Desalination 183 (2005) 167–172	In this work we present a new type of solar integrated-collector-storage system used in agricultural greenhouses.	Tunisia	NA	Solar energy	WT/domestic	The use of the storage collector allows the simplification of the current system and the use of a heat exchanger having a great heat transferring surface with a high effectiveness.
Bouguecha, S., Hamrouni, B., Dhahbi, M.	2005	Small scale desalination pilots powered by	Desalination 183 (2005) 151–165	This paper aims to identify the potential of renewable energies,	Tunisia	NA	RO	WT/Sea water	Economic feasibility has been demonstrated for MESS, but RO-PV and

(*continued*)

Authors	Year	Title	Journal name, volume (issue)	Abstract	Country	Funding Agent	Technology used	USE (WT/WWT)	Recommendations
		renewable energy sources: case studies		mainly solar and geothermal. It illustrates the quality of distributed water in rural communities.					MD-GW remain expensive solutions.
Fersi, C., Gzara, L., Dhahbi, M.	2005	Treatment of textile effluents by membrane technologies	Desalination 185 (2005) 399–409	This paper presents the results of the laboratory-scale membrane experiments of textile industry effluents previously biologically treated by activated sludge.	Tunisia	NA	MF, UF, NF	WWT/Industrial	Tunisian ministry of scientific research and competences development
Saddoud, A., Hassaïri, I., Sayadi, S.	2006	Anaerobic membrane reactor with phase separation for the treatment of cheese whey	Bioresource Technology 98 (2007) 2102–2108	This study examined the feasibility of applying an anaerobic membrane bioreactor with phase separation (acidogenesis/ methanogenesis) to treat cheese whey.	Tunisia	(MRSTDS, Tunisia), Mr A. Hajji (Engineering School of Sfax)	Cross-flow microfiltration	Cheese whey	The data clearly indicated that anaerobic treatment coupled to cross-flow microfiltration of cheese whey was successful.
Masmoudi, S., Larbot, A., El Feki, H., Ben Amar, R.	2006	Elaboration and properties of new ceramic microfiltration membranes from natural and synthesised apatite	Desalination 190 (2006) 89–103	In this work, we were interested in the development and the characterisation of new mineral microfiltration membranes coated on apatite macroporous supports which are prepared and characterised beforehand.	Tunisia	NA	Ceramic microfiltration membranes	WWT/Industrial	The results show that apatite is appropriate material for the development of microfiltration membranes applied to the industrial wastewater treatment. At the same time, a partial desalination can be obtained.
Yildiz, E., Keskinler, B., Pekdemir, T., Akay, G., Nuhŏglu, A.	2005	High strength wastewater treatment in a jet loop membrane bioreactor: kinetics and performance evaluation	Chemical Engineering Science 60 (2005) 1103–1116	The aim of this study is to determine the overall volumetric mass transfer coefficient (KLa), oxygen transfer efficiency, bio-kinetic coefficients based on	Turkey	NA	Membrane bioreactor (JLRs)	WWT/domestic	We can conclude that, due to its easy operation and simple construction, JL-MBRs have a high potential for use in the pre-treatment of high strength wastewater.

(*continued*)

Authors	Year	Title	Journal name, volume (issue)	Abstract	Country	Funding Agent	Technology used	USE (WT/WWT)	Recommendations
				both Monod equation, and SOUR from a laboratory scale JL-MBR wastewater treatment system.					
Ince, O., Kolukirik, M., Oz, N.A., Ince, B.K.	2005	Comparative evaluation of full-scale UASB reactors treating alcohol distillery wastewaters in terms of performance and methanogenic activity	Chem Technol Biotechnol 80: 138–144 (2005)	In this study, two full-scale up flow anaerobic sludge blanket (UASB) reactors, namely TUASB and CUASB, at the wastewater treatment plants of the Tekirdäg Alcohol (Raki) and Canakkale Alcohol (Cognac) distilleries were investigated in terms of performance, acetoclastic methanogenic capacity and microbial composition.	Turkey	Istanbul Technical University and Bogaziç University.	UASBR	WWT/Industrial	Despite all restrictions, direct microscopic counts provide general information about the numbers and morphology of the methanogenic species.
Cinar, O., Hasar, H., Kinaci, C.	2005	Modelling of submerged membrane bioreactor treating cheese whey wastewater by artificial neural network	Journal of Biotechnology 123 (2006) 204–209	A submerged membrane bioreactor receiving cheese whey was modelled by artificial neural network and its performance over a period of 100 days at different solids retention times was evaluated with this robust tool.	Turkey	Prof. Dr. A. Serdar Yılmaz and Assistant Prof. Dr. Abdülhamit Subaşin the Department of Electrical and Electronics Engineering of Kahramanmaraş Sütçü İmam University, Kahramanmaraş, Turkey	SMBR	WWT/cheese whey	This study demonstrates that artificial neural networks provide a robust tool for predicting the performance of membrane bioreactors even though such systems involve highly complex physical and biochemical mechanisms controlled by the membrane and the microorganisms.
Aybar, H.S.	2006	Mathematical modelling of an inclined solar water distillation system	Desalination 190 (2006) 63–70	An inclined solar water distillation (ISWD) system, which generates distilled water and hot water at the same time, was modelled and simulated.	Turkey	NA	Solar distillation	WT/SW	The system was simulated using actual deviations of solar intensity and environment temperature during a typical summer day. The system can generate 3.5–5.4kg/d distilled water. The

(continued)

Authors	Year	Title	Journal name, volume (issue)	Abstract	Country	Funding Agent	Technology used	USE (WT/WWT)	Recommendations
									temperature of the generated hot water is about 40C, which is good enough for domestic usage.
Yamall, C., Solmus, I.	2007	Theoretical investigation of a humidification/ dehumidification desalination system configured by a double-pass flat plate solar air heater	Desalination 205 (2007) 163–177	The aim of this study is to investigate theoretically the effect of different system operating conditions, types of air heater, and some different design parameters and a weather condition on a solar water desalination system performance under the climatological conditions of Ankara, Turkey.	Turkey	NA	Desalination	WT/SW	Theoretical results illustrated that the system productivity is strongly affected by the solar air heater area and, slightly influenced by the wind speed variations and the bottom heat loss coefficients of the solar air heater and water storage tank.
Peng, W., Escobar, I.C.	2005	Evaluation of Factors Influencing Membrane Performance	Environmental Progress (Vol. 24, No. 4) 2005, 392–399	The objective of this study was to investigate the fouling behaviours of different commercial membranes under various physical and chemical conditions to identify the dominant effects on the flux decline.	USA	The authors thank Dr. Mark Wilf (Hydranautics), Tom Stocker (Koch Membranes) and Dr. Seungkwan Hong (University of Central Florida).	NF & RO	WT/Industrial	It was found that the specific flux of TFC-S declined very quickly, ESPA1 specific flux decline rate decreased as time increased, and NTR7450 displayed a nearly stable specific flux.
Escobar, I.C.	2005	Membrane Developed Systems for Water and Wastewater Treatment	Environmental Progress (Vol. 24, No. 4) December 2005, 355–357	Existing water supplies may be limited in quantity or quality for meeting the increasing demands from population growth and industry expansion.	USA	NA	UF & RO	W&WWT	The next article in this special issue will show that fouling layers present on membranes in full-scale installations result in an improved rejection of solutes.

Annex II

© 2011 IWA Publishing. *Membrane Technology in Water Treatment in the Mediterranean Region (ProMembrane).* By Antonia Lorenzo and Aníbal Vega. ISBN: 9781843393702. Published by IWA Publishing, London, UK.

Authors	Year	Title	Journal name, volume (issue)	Abstract	Country	Funding Agent	Technology used	USE (WT/WWT)	Recommendations
Boerlage, S.F.E., Kennedy, M., Tarawneh, Z., Faber, R.D., Schippers, J.C.	2004	Development of the MFI-UF in constant flux filtration	Desalination 161 (2004) 103–113	The MFI- is determined in constant pressure filtration with the flux deceasing during the test. However, many membrane systems, e.g., reverse osmosis (RO), operate at constant flux with pressure increasing when fouling occurs. UF.	Australia	This research was jointly funded by Novem (Netherlands Agency for Energy and Environment) within the Subsidy Programme of New Technology for Industry (MINT), Kiwa N.V.Water Research and Hitma B.V.	MFI-UF	WT	The MFI-UF determined in constant flux filtration is expected to more closely simulate fouling at the surface of a membrane and improve the accuracy of fouling prediction by the MFI model. However, further investigation of the MFI-UF in the constant flux mode at lower applied flux, i.e., closer to RO operation, is required. Hence, in future work a reference pressure needs to be identified and a method developed to correct the fouling index (and MFI-UF) to the reference flux when conducted at non-referencec onditions.
Nghiem, L., Schafer, A., Elimelech, M.	2004	Removal of Natural Hormones by Nanofiltration Membranes: Measurement, Modelling, and Mechanisms	Environ. Sci. Technol. 2004, 38, 1888–1896	This study aimed at elucidating the removal mechanisms of natural steroid hormones by NF membranes.	Australia	US National Science Foundation, and the University of Wollongong and the Australian Institute of Nuclear Science and Engineering for a octoral scholarship to L.D. Nghiem.	NF	WT/Natural Hormones	Hormone diffusion in the membrane polymeric matrix most likely depends on the size of the hormone molecule, hydrogen bonding of hormones to membrane functional groups, and hydrophobic interactions of the hormone with the membrane polymeric matrix.
Schäfer A., Broeckmann, A., Richards, B.	2005	Membranes and renewable energy - a new era of sustainable development for developing countries	Membrane Technology November 2005	Details of a system that combines a solar energy system with hybrid ultrafiltration and reverse osmosis/ nanofiltration (RO/NF). The treatment accomplishes dual	Australia	Australian Research Council (ARC).	RO/NF	WT/BW	Further work is needed to integrate such treatment systems in a socially, economic and environmentally sustainable way.

(*continued*)

Authors	Year	Title	Journal name, volume (issue)	Abstract	Country	Funding Agent	Technology used	USE (WT/WWT)	Recommendations
				barrier disinfection, desalination and the removal of trace contaminants, such as arsenic.					
Nghiem, L., Schafer, A.I., Elimelech, M.	2005	Pharmaceutical Retention Mechanisms by Nanofiltration Membranes	Environ. Sci. Technol. 2005, 39, 7698–7705	The objective of this study is to examine the removal mechanisms of three pharmaceuticals sulfamethoxazole, carbamazepine, and ibuprofensby two thin-film composite NF membranes.	Australia	University of Wollongong. The Australian Institution of Nuclear Science and Engineering. Shimadzu (Sydney, Australia).	NF	WWT/ pharma-ceuticals	The interaction mechanism responsible for separation remains steric in nature. Because the average pore radius of the NF-90 membrane is substantially smaller than the radii of the three pharmaceuticals used in this study, no polarity effects can be observed with this tight NF-90 membrane, and the retention is nearly complete.
Masson, L., Richards, B.C., Schaifer, A.I.	2005	System design and performance testing of a hybrid membrane photovoltaic desalination system	Desalination 179 (2005) 51–59	Several parameters were examined in order to optimise the system performance, including i) feed water salt concentration, ii) operating pressure, iii) system recovery, iv) specific energy consumption (SEC) and v) salt retention.	Australia	Australian Research Council (ARC)	hybrid membrane photovoltaic, RO	WT/Surface water	Additional research is in progress to evaluate the long-term performance of the system (e.g. considering fouling and cleaning of the membranes), to optimise the energy efficiency, hence reduce initial costs and to assess the performance in terms of trace contaminants removal.
Chang, S., Fane, A.G., Waite, T.D.	2006	Analysis of constant permeate flow filtration using dead-end hollow fiber membranes	Journal of Membrane Science 268 (2006) 132–141	Dead-end filtration of colloids using hollow fibers has been analysed theoretically and experimentally.	Australia	Australia Research Council	hollow fiber membrane	WT	A mathematical model for dead-end hollow fiber membrane filtration under constant permeate flow condition using the Hagen–Poiseuille equation, the filtration

(*continued*)

Authors	Year	Title	Journal name, volume (issue)	Abstract	Country	Funding Agent	Technology used	USE (WT/WWT)	Recommendations
									equation, and the advanced cake filtration theory
Nghiem, L.D., Schäfer, A.I.	2006	Critical risk points of nanofiltration and reverse osmosis processes in water recycling applications	Desalination 187 (2006) 303–312	This study Focusing on the adsorption and release processes of EDCs during membrane filtration, and study highlights potential risks associated with the treatment, disposal of concentrate and membrane cleaning wastewater as well as provides orders of magnitudes of concentrations to be expected. Estimations are based on research investigations and assumptions realistic for large scale applications.	Australia	Australian Government's innovation statement Backing Australia's Ability, Commonwealth Department of Education Science and Training and Engineering (AINSE)	NF&RO	WWT/ Municipal	Membrane can serve as a large reservoir for EDCs and their release is likely during membrane cleaning or erratic pH variation during operation. Complete desorption of EDC to the membrane cleaning solution at high pH results in a high concentration of EDCs in the spent cleaning solution. Treatment of the concentrate and the spent cleaning solution need to be considered when EDCs are amongst the target contaminants in NF/RO applications.
Nghiem, L.D., Oschmann, N., Schäfera, A.I.	2006	Fouling in greywater recycling by direct ultrafiltration	Desalination 187 (2006) 283–290	The objectives of this study are to evaluate the feasibility of submerged ultrafiltration in greywater recycling under realistic concentration variations of common greywater constituents and to understand the impact of those constituents on fouling behaviour.	Australia	Gosford City Council, Commonwealth Department of Education Science and Training.	UF	WWT/ Greywater	Further work is in progress to evaluate the impact of other foulants such as proteins and surfactants, effects of foulants on retention of trace contaminants and the successful continuous operation and maintenance of larger scale units.

(*continued*)

Authors	Year	Title	Journal name, volume (issue)	Abstract	Country	Funding Agent	Technology used	USE (WT/WWT)	Recommendations
Schäfer, A., Broeckmann, A., Richards, B.	2005	Membranes and renewable energy - a new era of sustainable development for developing countries	Membrane Technology November 2005	Details of a system that combines a solar energy system with hybrid ultrafiltration and reverse osmosis/nanofiltration (RO/NF). The treatment accomplishes dual barrier disinfection, desalination and the removal of trace contaminants, such as arsenic.	Australia	Australian Research Council (ARC).	RO/NF	WT/BW	Further work is needed to integrate such treatment systems in a socially, economic and environmentally sustainable way.
Banasiak, L.J., Kruttschnitt, T.W., Schäfer, A.I.	2007	Desalination using electrodialysis as a function of voltage and salt concentration	Desalination 205 (2007) 38–46	In this study the performance of ED for the desalination of different salt solutions and the removal of fluoride and nitrate removal from a brackish groundwater was conducted. The influence of applied voltage and initial salt concentration of a solution was also determined.	Australia	The project is funded through the Australian Research Council Linkage Project in collaboration with Brisbane Water as well as the ARC Discovery Project. Eurodia Germany and France are thanked for the provision of membrane samples for this project, Berghof for the donation of the electrodialysis stack and Wytze Meindersma from the University of Twente, Netherlands and Bart van der Bruggen at the University of Leuven for useful discussions.	ED	WT/ brackish ground water	This study showed that desired drinking water can be obtained by the electrodialysis of ground water with fluoride and nitrate concentrations above Drinking Water Guidelines levels. Further work is required to determine fouling mechanisms as well as trace contaminant removal mechanisms.
Fuchs, W., Binder, H., Mavrias, G., Braun, R.	2003	Anaerobic treatment of wastewater with high organic content using a stirred tank reactor coupled with	Water Research 37 (2003) 902–908	In this paper, data from laboratory scale experiments are presented that show the advantages of this	Austria	NA	MBR	WWT/ Municipal	The membrane bioreactor is specifically useful for the treatment of complex wastewater rich in slowly degradable solids, which

(*continued*)

Authors	Year	Title	Journal name, volume (issue)	Abstract	Country	Funding Agent	Technology used	USE (WT/WWT)	Recommendations
		a membrane filtration unit		system in terms of volumetric loading rates, effluent quality and process stability.					are often treated insufficiently by other standard anaerobic treatment systems.
Schoeberl, P., Brik, M., Braun, R., Fuchs, W.	2004	Treatment and recycling of textile wastewater case study and development of a recycling concept	Desalination 171 (2004) 173–183	For a specific textile finishing company strategies for water recycling and recovery of valuable chemicals have been developed.	Austria	NA	MBR,UF	WWT/ Industrial	In conclusion, in regard to the investigated company, the process integrated recycling of washing fluids exhibits obvious advantages over end-of-pipe treatment.
Clara, M., Strenn, B., Gans, O., Martinez, E., Kreuzinger N., Kroiss, H.	2005	Removal of selected pharmaceuticals, fragrances and endocrine disrupting compounds in a membrane bioreactor and conventional wastewater treatment plants	Water Research 39 (2005) 4797–4807	Eight pharmaceuticals, two polycyclic musk fragrances and nine endocrine disrupting chemicals were analysed in several waste water treatment plants (WWTPs).	Austria	EU	MBR	WWT/ Municipal	Comparable effluent concentrations and removal rates for the different investigated substances were observed in the conventional WWTP and the membrane bioreactor.
Misra, B.M.	2007	Seawater desalination using nuclear heat/ electricity — Prospects and challenges	Desalination 205 (2007) 269–278	This paper presents the salient activities on nuclear desalination in the Agency and in the interested Member States. In particular, some data from the feasibility studies carried out in the two south Mediterranean countries — Egypt and Tunisia — for their nuclear power and desalination projects are presented.	Austria	NA	hybrid MED–RO.	WT/SW	The market, the nuclear power economics and its advantages in terms of reduced emissions will have a major impact on large-scale deployment of nuclear desalination.

(continued)

Authors	Year	Title	Journal name, volume (issue)	Abstract	Country	Funding Agent	Technology used	USE (WT/WWT)	Recommendations
Methnani, M.	2007	Influence of fuel costs on seawater desalination options	Desalination 205 (2007) 332–339	In this paper, we use a recently upgraded version of the IAEA Desalination Economic Evaluation Program (DEEP-3) to analyze the effects of these cost trends on reference estimates of seawater desalination costs.	Austria	NA	RO	WT/SW	The RO cost advantage compared to MED, would only be challenged at much higher salinities or increasingly poor quality of the feed water.
Al-Bastaki, N.M., Al-Qahtani, H.I.	1994	Assessment of effects on the reverse osmosis of salt/water solutions by using a spiral wound polyamide membrane	Desalination, 99 (1994) 159–168	The present work is concerned with the effect of varying the feed temperature on the water flux and salt rejection at various pressure levels.	Bahrain	NA	RO	WT/Salt water	The salt passage increased with increasing the temperature from about 12°C to 30°C. However, changing the pressure from 20 bars to 60 bars did not affect the salt passage for the low concentration solution and only resulted in a small decrease in the salt passage for the high concentration solution, particularly at higher temperatures.
Al-Bastaki, N.M.S.	1994	Behaviour of stepped GRP shells used in the reverse osmosis aerators for the water desalination process	Desalination, 94 (1994) 307–319	This work was aimed at studying the performance of shells which have been in operation for a period of 5 years, in order to assess their suitability to be reused after they have consumed their suggested lifetime.	Bahrain	NA	RO	WT/SW	Further study is required by examining tubes of different ages to determine whether there is a trend for deterioration in the properties with age, particularly in the region inside the circumferential groove where high axial strains were observed.
AL-Qahtany, H.I., AL-Bastaki, N.M.S.	1995	Effect of aging on the performance of RO hollow fibber membranes in a	Desalination 101 (1995) 177–183	In the present work we discuss the change in behaviour and efficiency of the hollow fibber	Bahrain	The data used in this study were supplied by the Ras Abu-Jarjur RO plant in Bahrain. The authors would like to	RO	WT/SW	It can be said that the overall performance of the studied block is within the expected range arid that the operating problems

(continued)

Authors	Year	Title	Journal name, volume (issue)	Abstract	Country	Funding Agent	Technology used	USE (WT/WWT)	Recommendations
		section of an RO plant		membranes used in the Ras Abu Jarjur RO plant in the State of Bahrain as a result of aging.		thank Mr. Mahmood AL-Arrayed, Mr. Khalid Al-Mansoor, and Mr. Hermoso for supplying the data and their support and help.			faced have been successfully handled and solved.
AL-Bastaki, N.M., Abbas, A.	1999	Improving the permeate flux by unsteady operation of a RO desalination unit	Desalination 123 (1999) 173–176	In a recent study by the authors of this work, periodic operation of a spiral-wound RO desalination unit was used using a symmetrical square wave pressure function.	Bahrain	NA	RO	WT/SW	The results have shown that the use of asymmetric operation improves the permeate flow rate for time split values in the range of 0.3 to 1, with the optimum value occurring at ~, = 0.6
AL-Bastaki, N.M., Abbas, A.	1999	Modelling an industrial reverse osmosis unit	Desalination 126 (1999) 33–39	A simplified model was used to predict the performance of hollow fibber RO membranes.	Bahrain	NA	RO	WWT/ Industrial	The theoretical analysis indicated that the pressure inside the fibber bore and the variation of concentration on the shell side of the fibber bundle significantly affected the results and, thus, cannot be ignored.
Abbas, A., AL-Bastaki, N.	2000	Flux enhancement of RO desalination processes	Desalination 132 (2000) 21–27	The main objective of the present work is the development of a simple model which may be used to predict the performance of RO systems under periodic operation.	Bahrain	NA	RO	WT/SW	The experimental results showed that operation of RO membranes in the unsteady-state mode using an asymmetric pressure square wave function may lead to improvements in both the permeation rates and the salt rejection of the system.
Al-Bastaki, N.M., Abbas, A.	2000	Predicting the performance of RO membranes	Desalination 132 (2000) 181–187	In the present work a comparative study was performed on the modelling of both	Bahrain	NA	RO	WT/SW	Prediction of the recovery can be significantly affected by the simplifying assumptions. Significant

(continued)

Authors	Year	Title	Journal name, volume (issue)	Abstract	Country	Funding Agent	Technology used	USE (WT/WWT)	Recommendations
				spiral-wound and hollow-fibber RO membranes.					over-estimation of the recovery for the hollow-fibber and spiral-wound membranes was found. For the spiral-wound membrane, ignoring the pressure drops was less significant.
Abbas, A., Al-Bastaki, N.	2001	Performance decline in brackish water FilmTec spiral wound RO membranes	Desalination 136 (2001) 281–286	In this paper, long-term operational data from a medium scale RO plant based on brackish water FilmTee spiral wound RO membranes are presented and analysed.	Bahrain	Mr. Seena Jaberi, General Manager of Al-Sane & Partners Soft Drinks Co.	RO	WT/BW	Using a low salinity ground water, the performance of spiral wound FilmTec RO membranes exhibited typical declines in water permeability and salt rejection.
Al-Bastaki, N., Abbas, A.	2001	Use of fluid instabilities to enhance membrane performance: a review	Desalination 136 (2001) 255–262	This paper presents a review of some of the key methods of generating flow instabilities that have been implemented in membrane separation processes by various workers during the past few years (mainly the 1990s).	Bahrain	NA	RO, UF, NF, MF	WT/SW	Several novel techniques of generating fluid instabilities for membrane processes have been developed and reported in the literature during the past few years.
Al-Baski, N., Abbas, A.	2003	Permeate recycle to improve the performance of a spiral-wound RO plant	Desalination 158 (2003) 119–126	This paper presents the results obtained from a simulation study on the effect of permeate recycle on the performance of a spiral-wound RO plant.	Bahrain	Mr. Seena Jaberi, General Manager of Al-Sane & Partners Soft Drinks Co.	RO	WT/SW	This method resulted in estimates of the mass transfer coefficient which are about one half of those corresponding to the simplified method that assumes an open channel and ignores the presence of the spacer.
Al-Bastaki, N.M.	2003	Treatment of synthetic industrial wastewater with UV/TiO, and RO	Desalination 156 (2003) 193–197	This paper describes the work done to study the efficiency of two types of processes in removing a model	Bahrain	NA	RO&UV/TiO_2	WWT/ Industrial	The UV/TiO, photocatalytic process showed a much higher efficiency than RO in removing of traces of benzene from the feed water.

(*continued*)

Authors	Year	Title	Journal name, volume (issue)	Abstract	Country	Funding Agent	Technology used	USE (WT/WWT)	Recommendations
		using benzene as a model hydrocarbon		hydrocarbon, namely, benzene, which was added to water at a concentration of 100 ppm.					
AL-Kawari, M., AL-Bastaki, N.	2004	Copper dust formation in a large scale MSF desalination plant	Desalination 166 (2004) 153–156	The objective of the current work is to present the problem of copper dust formation, which was observed in the relatively recently constructed and operated MSF units at A1-Hidd Power and Water plant.	Bahrain	Dr. Abdul Majeed AL-Awadi, the Undersecretary for the Ministry of Electricity and Water for the support and encouragement to conduct this work. Thanks are also due to Dr. Khalid Burashid, Engineer Ahmed Basheer, and Eng. Mohamed.	MSF	WT/SW	It is recommended that the effect of operating variables and aggressive gases on the copper dusting and VSC should be studied, including temperature, CO_2, air and pollutants, such as ammonia.
Al-Bastaki, N., Banat, F.	2004	Combining ultrafiltration and adsorption on bentonite in a one-step process for the treatment of coloured waters	Resources, Conservation and Recycling 41 (2004) 103–113	In the present work, adsorption onto bentonite is combined with ultrafiltration to examine the suitability of such a combined process in treating coloured wastewaters. Methylene blue, a basic dye, was used as a model compound.	Bahrain	NA	UF & Bentonite	WWT/ Industrial	The hybrid process utilizes the capabilities of bentonite to adsorb dyes and the capabilities of UF to remove colloids. The addition of bentonite to the feed solution was found to increase the rejection coefficient of methylene blue, but considerably deteriorate the permeate flux.
Al-Bastaki, N., Abbas, A.	2004	Long-term performance of an industrial water desalination plant	Chemical Engineering and Processing 43 (2004) 555–558	The long-term performance of a medium-scale industrial spiral wound reverse osmosis (RO) water desalination plant was studied. Operational data were analysed for a	Bahrain	The authors thank Mr. Seena Jaberi, General Manager of Al-Sane & Partners Soft Drinks Co., for his support and permission to publish this paper.	RO	WT/BW	A comparison of the results presented in this paper with those reported in previous studies shows that the performance of any RO plant depends strongly on the local conditions particularly the feed quality. This means that when designing new plants, the performance data of

(*continued*)

Authors	Year	Title	Journal name, volume (issue)	Abstract	Country	Funding Agent	Technology used	USE (WT/WWT)	Recommendations
				period of 1500 days as a basis for evaluating the performance variation with time.					existing plants operating under quite different conditions from those of the future plant should be treated with great care.
Al-Bastaki, N.M.	2004	Performance of advanced methods for treatment of wastewater:UV/ TiO_2, RO and UF	Chemical Engineering and Processing 43 (2004) 935–940	The objective of this work was to study the performance of small scale UV/TiO_2 photocatalytic pilot plant process in treating secondary and tertiary treated effluent from a wastewater treatment plant in Bahrain.	Bahrain	Tubli Pollution Control Center, Ministry of Works, Research Deanship, University of Bahrain.	UV/TiO_2, RO and UF	WWT/	The three advanced treatment processes, namely, UV/TiO_2, UF and RO, resulted in a complete removal of S. COD was reduced by 64% using the UV/TiO_2 process and by 86 and 50% with each of the RO and UF processes. The treated water production cost using the UV/TiO_2 process is expected to be higher than UF and BWRO due to the higher estimated power consumption. The specific power consumption for the UV/TiO_2 process was estimated to be about 4 kWh/m^3 which was estimated to be comparable to the SWRO but higher than the BWRO (about 1.02 kWh/ m^3) and cross flow UF (about 0.93 kWh/m^3). On the other hand, the UF and RO technologies produce a waste stream in which the microorganisms and other suspended and organic matter are concentrated and which contain the fouling-control chemicals which result in a continuous disposal problem. In comparison, the UV/TiO_2 produces no waste stream, and the main environmental

(*continued*)

Authors	Year	Title	Journal name, volume (issue)	Abstract	Country	Funding Agent	Technology used	USE (WT/WWT)	Recommendations
									concern is the disposal of the spent catalyst which may be needed once every few years.
Al-Bastaki, N.	2004	Removal of methyl orange dye and Na_2SO_4 salt from synthetic waste water using reverse osmosis	Chemical Engineering and Processing 43 (2004) 1561–1567	The objective of the current work was to study the effect of varying the Na_2SO_4 salt and methyl orange (MO) dye concentrations on the performance of a spiral wound polyamide membrane.	Bahrain	Research Deanship, University of Bahrain for the financial support for this work.	RO	WWT/ synthetic	The experimental results showed that increasing the dye concentration from 500 to 1000 ppm resulted in a decrease in the salt rejection for all of the operating pressures and for both feed salt concentrations of 5000 and 10,000 ppm. The percent dye removal for the 50,000 ppm salt solutions were slightly higher than those obtained with the 10,000 ppm salt solutions.
Burashi, K., Hussain, A.R.	2004	Seawater RO plant operation and maintenance experience: Addur desalination plant operation assessment	Desalination 165 (2004) 11–22	This paper attempts to describe the operational and mechanical problems encountered in the plant from the commissioning period till date and analyses them in view of before and after rehabilitation and sheds light on the reasons that lead to 'the rehabilitation - the production improvement works'.	Bahrain	NA	RO	WT/Sea water	Addur SWRO plant is expected to produce 5 million gallons/d of improved water quality, which will increase when the full fledge rehabilitation programme is implemented.
Abbas, A., Al-Bastaki, N.	2005	Modelling of an RO water desalination unit using neural networks	Chemical Engineering Journal 114 (2005) 139–143	In this paper, a feed forward neural network (NN) model is developed to predict the	Bahrain	NA	RO	WT/Sea water	The NN was also found to interpolate the data with good accuracy. As expected, however, the network did not produce

(*continued*)

Authors	Year	Title	Journal name, volume (issue)	Abstract	Country	Funding Agent	Technology used	USE (WT/WWT)	Recommendations
				performance of a reverse osmosis (RO) experimental setup, which uses a FilmTec SW30 membrane.					acceptable results when used for data extrapolation.
Lo, K.V., Liao, E.H. & Gao, Y.C.	1994	Anaerobic treatment of swine wastewater using hybrid UASB reactors	Bioresource Technology 47 (1994) 153–157	The study examined the reactor performance in terms of its treatment efficiency and methane production, and evaluated the effect which the absence of seeding in the start-up process had on the efficiency of reactor performance.	Canada	Ms Jet Blake.	UASB	swine waste water	Hybrid UASB reactors could be used for the treatment of screened swine manure.
Adris, A.M., Lim, C.J., Grace, J.R.	1994	The Fluidized bed membrane reactor system: A pilot scale experimental study	Chemical Engineering Science, Vol. 49. No. 24B, pp. 5833–5843, 1994	A pilot scale reforming plant was commissioned to study the new fluidized bed membrane reactor (FBMR) system for steam reforming of natural gas.	Canada	STDF/AGAR grant from the Science Council of British Columbia.	FBMR (fluidized bed membrane reactor)	natural gas	Experiments have confirmed that the FBMR system gives higher conversions for SMR reactions due to in situ continuous selective separation and removal of hydrogen from the reacting mixture. A multivariate analysis indicates that the reaction conversion increases strongly with temperature.
Cicek, N.	2003	A review of membrane bioreactors and their potential application in the treatment of agricultural wastewater	Canadian biosystem engineering 45 (2003), 37–49	This paper summarizes the potential applications of the MBR technology for the treatment of wastewater from agricultural sources.	Canada	NA	MBRs	WWT/ agricultural	The possibility of combining the removal of organic matter, nutrients, toxic chemicals, and biological organisms in one treatment system is certain to fuel future research and development in this emerging field.

(*continued*)

Authors	Year	Title	Journal name, volume (issue)	Abstract	Country	Funding Agent	Technology used	USE (WT/WWT)	Recommendations
Humphries, J.R., Wood, M.S.	2004	Reverse osmosis environmental remediation. Development and demonstration pilot project	Desalination 168 (2004) 177–184	The reverse osmosis environmental remediation pilot project is considered to have been a success.	Canada	NA	RO	WT/ Sea water	The CANDESAL advanced RO seawater desalination technology can be successfully extended to environmental remediation applications.
Meyer, N., Parker, W.J., Van Geel, P.J., Adiga, M.	2005	Development of an electro deionization process for removal of nitrate from drinking water Part 1: Single-species testing	Desalination 175 (2005) 153–165	The research described in this paper focused on the development of an emerging technology known as electro deionization (EDI) that combines ED with ion exchange.	Canada	Environmental Science and Technology Alliance of Canada and the Natural Science and Engineering Research Council of Canada.	EDI	WT/DW	The recycle experiments provided a reasonable estimate of the removal efficiency and energy consumption at various influent concentrations when compared with once-through experiments.
Wolf, P.H., Siverns, S., Monti, S.	2005	UF membranes for RO desalination pretreatment	Desalination 182 (2005) 293–300	Considering ultrafiltration (UF) membrane pretreatment; a technology enabling RO membrane units to provide decades of reliable high quality product water.	Canada	NA	UF&RO	WT/ Sea water	Major benefits include much shorter seawater intake lines, simple and efficient sand filter retrofits, higher RO membrane operating flux, less footprint and reduced downtime, maintenance and operating costs.
Nakhla, G., Lugowski, A., Patel, J., Rivest, V.	2006	Combined biological and membrane treatment of food-processing wastewater to achieve dry-ditch criteria: Pilot and full-scale performance	Bioresource Technology 97 (2006) 1–14	Bioreactors operated at low SRTs and in combination with ultrafiltration can still achieve superior effluent quality that may meet reuse criteria at reasonable cost.	Canada	Leamington Water Pollution Control Plant and Ontario Clean Water Agency (OCWA).	UF	WWT/ Industrial	For industrial wastewaters, the filterability of the mixed liquor is influenced by the soluble organics, and may be low, thus necessitating operation of bioreactors at low mixed liquor solids.
Mittal, G.S.	2006	Treatment of wastewater from abattoirs before land application—a review	Bioresource Technology 97 (2006) 1119–1135	Wastewater in abattoirs can be reduced by treatment of immersion chiller effluent by membrane filtration which can produce recyclable water.	Canada	Ontario Ministry of Environment, and Ontario Ministry of Agriculture and Food.	membrane filtration	WWT/ abattoir	The planning for abattoir waste minimization should incorporate the following options (EPA, 2000): (i) changing the processes or equipment, (ii) changing the composition, packaging or durability of products,

(continued)

Authors	Year	Title	Journal name, volume (issue)	Abstract	Country	Funding Agent	Technology used	USE (WT/WWT)	Recommendations
									(iii) changing or reducing raw material inputs, (iv) improving the control of the process, (v) improving the materials handling and cleaning operations, (vi) improving the maintenance and repair of equipment, and (vii) recycling waste internally, re-using waste on site.
Yang, W., Cicek, N., John Ilg, J.	2006	State-of-the-art of membrane bioreactors: Worldwide research and commercial applications in North America	Journal of Membrane Science 270 (2006) 201–211	This paper aims to critically characterize and review worldwide academic research efforts in the area of MBRs as well as focus attention to commercial MBR applications in North America.	Canada	NA	MBR	WWT/ Industrial	it is expected that a significant increase in MBR plant capacity and widening of application areas will occur in the future.
Sorin, M.S., Jedrzejak, S., Bouchard, C.	2006	On maximum power of reverse osmosis separation processes	Desalination 190 (2006) 212–220	This paper considers the application of finite time thermodynamics (FTT) to reverse osmosis (RO) processes.	Canada	NA	RO	WT/ Sea water	This paper was a first attempt in applying FTT to reverse osmosis. Some phenomena were intentionally neglected and some assumptions were made about salt separation and power recovery efficiencies.
Wen C., Huang, X., Qian, Y.	1999	Domestic wastewater treatment using an anaerobic bioreactor coupled with membrane filtration	Process Biochemistry 35 (1999) 335–340	COD mass balance analysis showed that about half of the influent COD was converted to methane gas.	China	NA	membrane-coupled anaerobic bioreactor.	WWT/ Domestic	COD mass balance analysis showed that about half of the influent COD was converted into methane gas.
Zhang, S. Houten, R.V., Eikelboom, D.H., Doddema, H., Jiang, Z., Fan, Y., Wang, J.	2003	Sewage treatment by a low energy membrane bioreactor	Bioresource Technology 90 (2003) 185–192	The resulting high-quality and disinfected effluent implies that MBR processes can be especially suitable for reuse and recycling of wastewater.	China	TNO, RCEES&CAS and Royal Dutch Academy of Sciences.	MBR	WWT/ municipal	The transverse flow module is an effective type of membrane module and needs more attention in order to further optimize energy saving.

(*continued*)

Authors	Year	Title	Journal name, volume (issue)	Abstract	Country	Funding Agent	Technology used	USE (WT/WWT)	Recommendations
Gao, M., Yang, M., Li, H., Wang, Y., Pan, F.	2004	Nitrification and sludge characteristics in a submerged membrane bioreactor on synthetic inorganic wastewater	Desalination 170 (2004) 177–185	The purpose of this study is to elucidate nitrification performance and some microbiological characteristics of sludge in SMBR treating of inorganic wastewater over a long operational period.	China	Chinese Academy of Sciences (Hundreds-Talent Program) and the Key Laboratory of Environmental Science and Technology of Higher Education of Henan Province.	SMBR	WWT/ Industrial.	The present results show that SMBR can be operated efficiently and with stability as a high rate nitrifying technology that can be applied in the nitrification of wastewater containing a high concentration of ammonia nitrogen.
Wen, X., Ding, H., Huang, X., Liu, R.	2004	Treatment of hospital wastewater using a submerged membrane bioreactor	Process Biochemistry 39 (2004) 1427–1431	Performance of a submerged hollow fiber membrane bioreactor (MBR) for treatment of hospital wastewater was investigated.	China	NA	SMBR	hospital waste-water	A full-scale submerged membrane bioreactor was used effectively for hospital wastewater treatment. Membrane separation played an important role in ensuring excellent and stable effluent quality. The system had very high NH4+–N removal efficiency. Sludge production was low and table. Transmembrane pressure increased slowly during the operation period indicating that the start-up and operational strategy of the system was successful.
Gao, M., Yang, M., Li, H., Yang, Q., Zhang, Y.	2004	Comparison between a submerged membrane bioreactor and a conventional activated sludge system on treating ammonia-bearing inorganic wastewater	Journal of Biotechnology 108 (2004) 265–269	A submerged membrane bioreactor (SMBR) and a conventional activated sludge system (CAS) were compared in parallel over a period of 210 days on treating synthetic ammonia-bearing inorganic wastewater under similar conditions.	China	National Natural Science Foundation of China and State Hi-Tech Research and Development Project of the Ministry of Science and Technology, China.	SMBR	inorganic waste-water	The SMBR, which contained larger numbers of nitrifiers, was more effective and stable than the CAS in treating synthetic ammonia-bearing inorganic wastewater.

(*continued*)

Authors	Year	Title	Journal name, volume (issue)	Abstract	Country	Funding Agent	Technology used	USE (WT/WWT)	Recommendations
Ding, Z., Liu, L., EL-Bourawi, M.S., Ma, R.	2005	Analysis of a solar-powered membrane distillation system	Desalination 172 (2005) 27–40	In dry and rural areas, solar-powered membrane distillation (SPMD) technology is considered a feasible means for the production of pure water from brackish water.	China	China National Natural Science Foundation.	solar-powered membrane distillation (SPMD)	WT/BW	The SPMD plant shows that its performance is remarkably influenced by the amount of energy transported to the feed. Plant capacity is proportional to membrane area may not be always true. The effectiveness of improving plant capacity by increasing the membrane area largely depends on heat recovery. The MD starting temperature, t] t ~, has almost no relation to plant capacity. A higher tf ~ t" means less running time of the pumps, which is what we prefer.
Zhallg, J.C., Wang, Y.H., Song, L.F., Hu, J.Y., Ong, S.L., Ng, W.J., Lee, L.Y.	2005	Feasibility investigation of refinery wastewater treatment by combination of PACs and coagulant with ultrafiltration	Desalination 174 (2005) 247–256	The aim of the present work was to determine the influence of PACs and coagulant addition on the performance of the UF process.	China	NA	UF	WWT/ Industrial	It showed that the UF unit performance could be obviously improved by simultaneously adding PACs-2 and HCA into the system. The suitable amounts for PACs-2 and HCA were 15mg.L^{-1} and 0.8ml.L^{-1} respectively, in which the removal rates of TOCs in the wastewater were over 99%.
Zhang, L., Xie, L., Chen, H.-L., Gao, C.-J.	2005	Progress and prospects of seawater desalination in China	Desalination 182 (2005) 13–18	In this paper, the status of water resources in China has been analyzed, seawater desalination technologies used in China are presented	China	The National Basic Research Program of China.	RO	WT/ Sea water	In MSF, MED and SWRO processes for seawater desalination, the cost of SWRO is lower, but MSF or MED is more effective in some cogeneration plants in coastal region of China.

(*continued*)

Authors	Year	Title	Journal name, volume (issue)	Abstract	Country	Funding Agent	Technology used	USE (WT/WWT)	Recommendations
				and their economics are evaluated.					
Wang, D.-X., Su, M., Yu, Z.-Y., Wang, X.-L., Ando, M., Shintani, T.	2005	Separation performance of a nanofiltration membrane influenced by species and concentration of ions	Desalination 175 (2005) 219–225	Nanofiltration (NF), which has been largely developed over the past decade, is a promising technology for the treatment of organic and inorganic pollutants in surface and ground waters.	China	National Basic Research Program of China, Nitto-Denko Corporation.	NF	WT/ Surface & ground water.	The electrostatic effect is the major factor at low salt concentration which influenced the separation capability of the NF membrane. Also, the steric-hindrance effect is the major factor at high salt concentrations.
He, Y., Xu, P., Li, C., Zhang, B.	2005	High-concentration food wastewater treatment byan anaerobic membrane bioreactor	Water Research 39 (2005) 4110–4118	The viability of treating high-concentration food wastewater byan anaerobic membrane bioreactor (AMBR) was studied using polyethersulfone (PES) ultrafiltration membranes PES200, PES300, PES500 and PES700 with norminal molecular weight cutoff (MWCO) ranging from 20,000 to 70,000 Da.	China	National Science Foundation Commission of China.	AMBR	WWT/ Industrial	More frequent back-washing and efficient chemical cleaning should be used to control the biofouling layer and maintain an acceptable flux.
Liu, R., Huang, X., Xi, J., Qian, Y.	2005	Microbial behaviour in a membrane bioreactor with complete sludge retention	Process Biochemistry 40 (2005) 3165–3170	Long-term microbial behaviour was investigated in a membrane bioreactor with complete sludge retention.	China	NA	MBR	Sludge	Supernatant TOC slightly accumulated in the reactor. Inorganic substances slightly accumulated in the reactor. The biomass concentration levelled off at 16 g-MLSS l 1after 200 days of operation. The organic removal performance remained stably excellent whereas

(*continued*)

Authors	Year	Title	Journal name, volume (issue)	Abstract	Country	Funding Agent	Technology used	USE (WT/WWT)	Recommendations
									the membrane permeability of the mixed liquor decreased with the elapsed time.
Ren, N., Chen, Z., Wang, X., Hu, D., Wang, A.	2005	Optimized operational parameters of a pilot scale membrane bioreactor for high-strength organic wastewater treatment	International Biodeterioration & Biodegradation 56 (2005) 216–223	A pilot-scale test was conducted in a submerged membrane bioreactor (SMBR) for 452 days to treat high-strength traditional Chinese medicine wastewater from two-phase anaerobic digest effluent.	China	Research Center of Environmental Biotechnology in Harbin Institute of Technology.	SMBR	Traditional Chinese medicine wastewater	The simulation model can be used to control the operational parameters for obtaining the best effluent quality of the SMBR treatments.
Lv, W., Zheng, X., Yang, M., Zhang, Y., Liu, Y., Liu, J.	2006	Virus removal performance and mechanism of a submerged membrane bioreactor	Process Biochemistry 41 (2006) 299–30	Virus removal performance and mechanism of a SMBR were investigated by employing phage T4 as a model virus.	China	National Science Foundation of China.	SMBR	WWT/ municipal	In conclusion, data in this study showed that the well running SMBR had a good ability to intercept/inactivate the virus that surged suddenly into municipal wastewater. In order to preclude any virus leakage, the membrane module, with a pore size slightly smaller than diameter of the target virus, would be a safe option.
Fan, Y., Li, G., Wu, L., Yang, W., Dong, C., Xu, H., Fan, W.	2006	Treatment and reuse of toilet wastewater by an airlift external circulation membrane bioreactor	Process Biochemistry (2006)	Membrane bioreactor (MBR) is a high efficient technology for toilet wastewater treatment and reuse.	China	State Hi-Tech Research and Development Project of the Ministry of Science and Technology.	MBR	Toilet WWT	It presents that AEC-MBR is a very practical and economic technology for toilet wastewater reuse.
Kurniawan, T.A., Chana, G.Y.S., Loa, W.H., Babel, S.	2006	Physico–chemical treatment techniques for	Chemical Engineering Journal 118 (2006) 83–98	This article reviews the technical applicability of various physico–chemical	China	The Hong Kong Polytechnic University for the International Postgraduate Study	physico–chemical techniques for	WWT/ Industrial	Technical applicability, plant simplicity and cost-effectiveness are the key factors that play major roles

(*continued*)

Authors	Year	Title	Journal name, volume (issue)	Abstract	Country	Funding Agent	Technology used	USE (WT/WWT)	Recommendations
		wastewater laden with heavy metals		treatments for the removal of heavy metals such as Cd(II), Cr(III), Cr(VI), Cu(II), Ni(II) and Zn(II) from contaminated wastewater.		Scholarship, PolyU Research, Innovative Technology Commission, Research Grant Council (RGC) of Hong Kong.	electroplating wastewater.		in the selection of the most suitable treatment system for inorganic effluent.
Yun, Y., Ma, R., Zhang, W., Fane, A.G., Li, J.	2006	Direct contact membrane distillation mechanism for high concentration NaCl solutions	Desalination 188 (2006) 251–262	In this work, direct contact membrane distillation mechanism for high concentration NaCl solutions is studied. Based on this studying, the coefficients of water recovery are improved and energy consumption per ton fresh water is reduced, which provides a reference for DCMD producing fresh water from seawater.	China	Australia–China Institutional Links Program and the UNESCO Center.	DCMD& RO.	WT/ NaCL Sol	The results under the experimental conditions show that (1) water fluxes decrease as concentrations of NaCl solutions increase; (2) when concentrations of NaCl solutions are above 25%, water fluxes decrease sharply; (3) when concentrations of NaCl solutions are up to saturation, water fluxes gradually approach steady levels.
Yang, Q., Chen, J., Zhang, F.	2006	Membrane fouling control in a submerged membrane bioreactor with porous, flexible suspended carriers	Desalination 189 (2006) 292–302	The aim of this study was to control membrane fouling in terms of improving the characteristics of activated sludge and enhancing cake removal.	China	Shanghai Municipal Science & Technology Commission of China.	MBR	WWT/ domestic	The effect of suspended carriers on membrane permeability and fouling in submerged MBR and HMBR was investigated. Compared with MBR, the critical flux increased by about 20% and the cake resistance decreased by 86% in HMBR. During the long-term experiments, the increase rate of suction pressure for HMBR accounted for 30% of that for MBR.

(*continued*)

Authors	Year	Title	Journal name, volume (issue)	Abstract	Country	Funding Agent	Technology used	USE (WT/WWT)	Recommendations
Zhang, H.-M., Xiao, J.-N., Cheng, Y.-J., Liu, L.-F., Zhang, X-W., Yang, F.-L.	2006	Comparison between a sequencing batch membrane bioreactor and a conventional membrane bioreactor	Process Biochemistry 41 (2006) 87–95	An operation of a membrane bioreactor in sequencing batch mode named a sequencing batch membrane bioreactor (SBMBR) was investigated for enhancing nitrogen and phosphorus removal.	China	Chinese National Science Foundation Committee.	SBMBR,CMBR.	WWT/ Municipal	Both SBMBR and CMBR showed good performance on organic substance removal during the whole experimental period. The SBMBR demonstrated a good performance on nitrogen and phosphorus removal at different influent COD/TN ratios. The variation of influent COD/TN ratio affected the CMBR more extensively than the SBMBR. Specific nitrification rate tests demonstrated that the pH value could affect the activity of nitrifying bacteria with no irreversible harm. The sequencing batch operation mode can reduce membrane fouling although a higher EPS existed in the SBMBR system.
Yang, L., Shen, S.	2007	Assessment of energy requirement for water production at dual-purpose plants in China	Desalination 205 (2007) 214–223	This paper presents an economic analysis of the energy cost of water production for dual-purpose plants by the equivalent enthalpy drop theory (EEDT) and the off-design analytical theory on steam turbines.	China	Science and Technology Foundation of Liaoning, China.	LT–MED.	WT/SW	The results indicate that EECR is more accurate to be used to evaluate the economics of the desalination process than GOR that is widely used. They also show that EECR that is the function of the equivalent enthalpy drop.
Gao, P., Zhang, L., Cheng, K., Zhang, H.	2007	A new approach to performance analysis of a seawater desalination system	Desalination 205 (2007) 147–155	The object of modelling is a desalination set-up which was made by the Institute of Air-Conditioning and	China	Chinese Foundation Committee of Nature and Science.	Desalination	WT/SW	Using ANN, we conveniently obtained the real-time water production ratio and could adjust the water production parameters to make the

(*continued*)

Authors	Year	Title	Journal name, volume (issue)	Abstract	Country	Funding Agent	Technology used	USE (WT/WWT)	Recommendations
		by an artificial neural network		Solar Energy of Northwestern Polytechnical University.					system more efficient and to gain more fresh water. Overall results show that the ANN can be used as an alternative way to study the problem of the desalination.
Yuan, G., Zhang, H.	2007	Mathematical modeling of a closed circulation solar desalination unit with humidification–dehumidification	Desalination 205 (2007) 156–162	This study focuses on studying and analyzing the collector area, cooling water flow rate, seawater flow rate and structural dimension effect on fresh water production.	China	NA	Desalination	WT/SW	For a fixed solar area, the feed water flow rate increasing will decrease fresh water production because of the spray temperature decrease. The matching of the cooling water flow rate with the collector area is very important to obtain maximum fresh water production. The system can be operated round-the-clock it will improve fresh water production.
Tam, L.S., Tang, T.W., Lau, G.N., Sharma, K.R., Chen, G.H.	2007	A pilot study for wastewater reclamation and reuse with MBR/RO and MF/RO systems	Desalination 202 (2007) 106–113	A pilot study was conducted to evaluate the quality of the effluent of an enhanced tertiary treatment process consisting of MBR/RO or MF/RO units and explore the feasibility of reclamation of treated effluent for potable and non-potable reuse applications.	China	Drainage Services Department of the HKSAR Government and the Mitsubishi Rayon Co., Ltd. and Nitto Denko.	MBR/RO and MF/RO	WWT/ domestic	The results of this pilot trial illustrated that the MBR/RO and MF/RO technologies can be used to treat raw sewage and secondary effluent, respectively to produce reclaimable water.
Liu, Z., Qun, M., An, W., Sun, Z.	2007	An application of membrane bio-reactor process for the wastewater treatment of	Desalination 202 (2007) 144–149	This paper describes the design of the membrane bio-reactor (MBR) used as a backbone treatment	China	NA	membrane bio-reactor (MBR)	WWT/ domestic	The pretreatment of wastewater should be strengthened in order to reduce the membrane fouling and harmfulness

(*continued*)

Authors	Year	Title	Journal name, volume (issue)	Abstract	Country	Funding Agent	Technology used	USE (WT/WWT)	Recommendations
		Qingdao International Airport		component at the Qingdao International Airport Wastewater Treatment Station (WWTS) Qingdao, China, to biologically remove COD, nitrogen and phosphorus as well as the operating results of the station.					caused by the fine fiber and hair in the wastewater.
You, S.-H., Tseng, D.-H., Hsu, W.-C.	2007	Effect and mechanism of ultrafiltration membrane fouling removal by ozonation	Desalination 202 (2007) 224–230	This study is aimed using ozonation to remove the fouling that occurs on ultrafiltration membranes.	China	National Science Council, Republic of China.	UF	WWT/ Industrial	Ozone could inhibit fouling of the ultrafiltration membrane due to the presence of organic carbon as well as change the chemical structure and the characteristics of the filtration cake produced by the ultrafiltration of organic wastewater.
Chan, G.Y.S., Chang, J., Kurniawan, T.A., Fu, C.-X., Jiang, H., Je, Y.	2007	Removal of non-biodegradable compounds from stabilized leachate using VSEPRO membrane filtration	Desalination 202 (2007) 310–317	In this paper, the technical applicability of the vibratory shear-enhanced processing reverse osmosis (VSEPRO) membrane system for the treatment of stabilized leachate from the NENT landfill in Hong Kong was studied using both 99 and 96% NaCl rejection membranes.	China	Hong Kong Polytechnic University.	VSEPRO membrane filtration.	WWT/ Municipal	VSEPRO is technically applicable and appealing for the treatment of stabilized leachate. No biological process is required for further treatment of the leachate, thus reducing operational costs.
Yamagiwa, K., Ohmae, Y., Dahlan, M.H., Ohkawa, A.	1991	Activated Sludge Treatment of Small-Scale Wastewater by a Plunging Liquid Jet Bioreactor with Cross-Flow Filtration	Bioresource Technology 37 (1991) 215–222	A new energy-saving membrane bioreactor was Cf developed and applied to biological treatment of wastewater.	Japan	Grants-in-Aid for Scientific Research 'Formulation and Management of Man-Environmental System'/ Japan &TDK.	membrane bioreactor.	WWT/ Domestic	the present bioreactor system has considerable potential for small-scale biological treatment of wastewater such as livestock and domestic wastewater.

(*continued*)

Authors	Year	Title	Journal name, volume (issue)	Abstract	Country	Funding Agent	Technology used	USE (WT/WWT)	Recommendations
Murakami, M.	1994	Hydro-powered reverse osmosis (RO) desalination for co-generation: A Middle East case study	Desalinisation, 97 (1994) 301–311	A new co-generation application of the hydro-power development with application of reverse osmosis (RO) desalination was demonstrated.	Japan	NA	RO	WT/BW	This study attempts to evaluate some new non-conventional approaches to water resources which need to taken into account in building the new peace of the Middle East.
Yamagiwa, K., Oohira, Y., Ohkawa, A.	1995	Simultaneous removal of carbonaceous and nitrogenous pollutants by a plunging liquid jet bioreactor with crossflow filtration operated under intermittent aeration	Bioresource Technology 53 (1995) 57–62	The bioreactor system with the biomass carrier operated under intermittent aeration was judged to be excellent for simultaneous removal of TOC and nitrogen in small-scale treatment of domestic wastewater.	Japan	Grants-in-Aid for Scientific Research 'Formulation and Management of Man-Environmental System' from the Ministry of Education, Culture and Science, Japan. TDK Corporation for supplying the membrane module.	Membrane bioreactor, MF	WWT/ Domestic	Simultaneous removal of carbonaceous and nitrogenous pollutants can be carried out successfully in a bioreactor with the biomass carrier operated under intermittent aeration.
Ohya, H., Kim, J.J., Chinen, A., Aihara, M., Semenova, S.I., Negishi, Y., Mori, O., Yasuda. M.	1998	Effects of pore size on separation mechanisms of microfiltration of oily water, using porous glass tubular membrane	Journal of Membrane Science 145 (1998) 1–14	The effects of pore size on permeate flux and oil rejection were investigated and the filtration mechanisms were analyzed based on various types of filtration models.	Japan	Foundation of River and Watershed Environment Management, Japan and Environmental Conservation Project Fund of Kanagawa Prefecture, Japan.	MF	WWT/Oil	The transition time of separation mechanism was of the same value regardless of the membrane pore size. According to the size of the membrane pores, the blocking filtration mechanisms are different. The performance of microporous glass membranes was superior in the separation of oil from oily water.
Bian R., Yamamoto, K., Watanabe, Y.	2000	The effect of shear rate on controlling the concentration polarization and membrane fouling	the Conference on Membranes in Drinking and Industrial Water Production, Volume 1, pages 421–432	This paper deals with the results from experiments concerning the effects of an increasing shear rate at the edge of the membrane on	Japan	Core Research for Evolutionary Science and Technology (CREST) of Japan Science and Technology Corporation (JST) and	NF	WT/ Surface water	Concentration polarization not only results in a decrease of the removal of dissolved materials, but also in increase of filtration resistances.

(continued)

Authors	Year	Title	Journal name, volume (issue)	Abstract	Country	Funding Agent	Technology used	USE (WT/WWT)	Recommendations
				reducing concentration polarization and on controlling membrane fouling during filtration of river water.		by the Advanced Aqua Clean Technology 21st Centenary (ACT21) research project of the Japanese Ministry of Health and Welfare.			
Urase, T., Kagawa, C., Kikuta, T.	2005	Factors affecting removal of pharmaceutical substances and estrogens in membrane separation bioreactors	Desalination 178 (2005) 107–113	In this study, a laboratory-scale MBR was operated with artificial wastewater containing pharmaceuticals and estrogens, and factors influencing removal were investigated.	Japan	NA	Membrane separation bioreactor, M.F	WWT/ Industrial	Higher removal of acidic pharmaceutical substances was obtained in the case of lower pH operation because of the increased tendency of adsorption to the sludge particles.
Kimura, K., Hara, H., Watanabe, Y.	2005	Removal of pharmaceutical compounds by submerged membrane bioreactors (MBRs)	Desalination 178 (2005) 135–140	Membrane bioreactors (MBRs) have gained significant popularity as an advanced wastewater treatment technology and might be effective in removing such organic micro-pollutants.	Japan	NA	MBR	WWT/ Industrial	MBRs can remove PhACs with complicated structures that cannot be sufficiently removed by CAS,However, there was no significant difference between MBR and CAS when PhACs contained chlorine in their structure.
Jung, Y.J., Kiso, Y., Yamada,T., Shibata, T., Lee, T.G.	2006	Chemical cleaning of reverse osmosis membranes used for treating wastewater from a rolling mill process	Desalination 190 (2006) 181–188	The fouling materials obtained from exhausted RO membranes, used to treat wastewater from a rolling mill process in the steel industry, were analyzed.	Japan	Grant-in-Aid for Scientific Research, JSPS, COE Program.	RO	WWT/ Industrial	The results of the surface observation suggest that the trace organic foulants acted as a binder for inorganic foulants. Organic contaminants were present at a very low level. The EGMB solution provided effective recovery in spite of the result that major fouling materials were inorganic substances. The surface of the EGMB cleaning membrane was similar to that of the virgin one.

(*continued*)

Authors	Year	Title	Journal name, volume (issue)	Abstract	Country	Funding Agent	Technology used	USE (WT/WWT)	Recommendations
Collins, J.H., Yoon, S.H., Musale, D., Kong, J.F., Koppes, J., Sundararajan, S., Tsai, S.P., Hallsby, G.A., Cachia, P., Kronoveter, K.	2006	Membrane performance enhancer evaluations on pilot- and full-scale membrane bioreactors	Water and Environment Journal 20 (2006) 43–47	In this study, MPE50 was tested in full-scale MBR plants to confirm the results obtained with the lab and pilot experiments. A method to determine the initial dosage will be demonstrated in addition to the pilot- and full-scale results.	Japan	The authors thank Mr. Hoon Kim of Genix Engineering, Korea, for his help with the field trial using Yuasa membranes, Dennis Livingston of Enviroquip Incorporated, USA, for his help with the pilot and full-scale tests and Trevor Miller and Lyle Ferguson of the Running Springs County Water District, Running Springs, CA for their help with the full-scale test.	MBRs	WWT/ MPE50	Initial MPE50 dosage can be determined on the basis of visual jar tests, in which different amounts of MPE50 are added to the mixed liquor, and particle size and supernatant clarity are monitored visually. An additional 50% of MPE50 should be added to the bioreactor, because of the delayed diffusion of SMP into the mixed liquor. Maximum peak flow could be obtained without any significant membrane fouling as evidenced by the low TMP and the subsequent return to normal TMP after the peak flow events were finished. Long-term, full-scale field trials showed 72% flux increases. MPE50 is found to be beneficial in reducing foam in addition to fouling control.
Xing, C.-H., Yamamoto, K., Fukushi, K.	2006	Performance of an inclined-plate membrane bioreactor at zero excess sludge discharge	Journal of Membrane Science 275 (2006) 175–186	The objective of this paper is to introduce an inclined-plate membrane bioreactor (iPMBR) to challenge zero excess sludge discharge at continuous removals of both carbonous and nitrous nutrients from municipal wastewater.	Japan	Japan Society of Promotion of Science for the financial support and to Mitsubishi Rayon Co. Ltd. for their supply of membrane modules.	iPMBR	WWT/ Municipal	Future research may be required to focus on the mathematical understanding of high MLSS sedimentation/ separation in the presence of intermittent air blowing and the development of design guidelines for the iPMBR process.

(*continued*)

Authors	Year	Title	Journal name, volume (issue)	Abstract	Country	Funding Agent	Technology used	USE (WT/WWT)	Recommendations
Chae, S.R., Kang, S.T., Lee, S.M., Lee E.S., Oh, S.E., Watanabe, Y., Shin, H.S.	2007	High reuse potential of effluent from an innovative vertical submerged membrane bioreactor treating municipal wastewater	Desalination 202 (2007) 83–89	The main objective of this study is to investigate the reuse possibility of effluent from a vertical submerged membrane bioreactor (VSMBR) treating municipal wastewater.	Japan	Ministry of Environment.	MBR	WWT/ Municipal	This process can be strongly recommended for the advanced wastewater treatment and water reuse purposes in urban and rural area.
Urase, T., Satoa, K.	2007	The effect of deterioration of nanofiltration membrane on retention of pharmaceuticals	Desalination 202 (2007) 385–391	The retention of 8 acidic and 2 neutral pharmaceuticals in nanofiltration (NF) was investigated by a loose NF membrane and a tight NF membrane.	Japan	NA	NF	WWT/ Industrial	The retention of pharmaceuticals was affected by the electric repulsion effect in addition to the size exclusion effect.
Jung, Y.J., Kiso, Y., Park, H.J., Nishioka, K., Min, K.S.	2007	Rejection properties of NF membranes for alkylphenols	Desalination 202 (2007) 278–285	In this study. They examined the removal performance of NF membrane for APs from the following two points of view: (1) the rejection properties of four types of NF membranes for alkylphenols (APs) were examined experimentally with a batch-type apparatus, and (2) the adsorption properties of the APs on NF membranes were also investigated by the batch-type adsorption experiments and an inverse HPLC technique.	Japan	Grant-in-Aid for Scientific Research from the Ministry of Education, Culture, Sports, Science and Technology of Japan and by the 21st Century COE Program.	NF	WT/AP	The highest desalting membrane (Memb-1) rejected effectively 4-n-octylphenol and 4-n-nonylphenol, but the other membranes did not show any effective removal performance for the APs. (2) The rejection degree was correlated better with molecular weight of the APs than with the log P. (3) The APs were adsorbed on the membrane polymer and the adsorption was mainly controlled by the hydrophobic interaction.

(*continued*)

Authors	Year	Title	Journal name, volume (issue)	Abstract	Country	Funding Agent	Technology used	USE (WT/WWT)	Recommendations
Jolanta Bohdziewicz.	1998	Biodegradation of phenol by enzymes from Pseudomonas sp. immobilized onto ultrafiltration membrane	Process Biochemistry Vol. 33. No. 8, pp. 811–818, 1998	The immobilized enzyme system was applied to the biodegradation of phc not in coke wastewaters.	Poland	Prof. S. I_ abuzek, research grant from the Polish Committee for Scientific Research.	UF/enzymatic membrane.	WWT/ Domestic	Phenol biodegradation by enzymes isolated from a bacterial strain of Pseudomonas and permanently immobilized in the porous structure of an asymmetric polyacrylonitrile ultrafiltration membrane.
Kujawski, W., Krajewski, S.R.	2004	Sweeping gas pervaporation with hollow-fibber ion-exchange membranes	Desalination 162 (2004) 129–135	The selective and transport properties of an anion exchange hollow-fibber membrane determined in the sweeping gas pervaporation process used for the dehydration of different organic solvents.	Poland	R. Wachs and Dr. E. Korngold, Anna Gil and Mr. Andrzej Bardonski.	Pervasive hollow-fibber ion-exchange membrane.	water-organic mixtures	Selectivity and transport results obtained for the Pervasive ion-exchange membrane in contact with different water-organic mixtures are in a good accordance with other results on pervaporation with the ionomeric membranes.
Turek, M., Dydo, P., Was, J.	2007	High efficiency electrodialysis reversal of concentrated calcium sulphate and calcium carbonate solutions	Desalination 205 (2007) 62–66	The objective was to assess the possibility of EDR system efficiency enhancement by ED stack modification.	Poland	Polish Ministry of Science and Information Technology, and the SUT project.	Electrodialysis reversal.	WT	It was also shown that controlled gypsum crystallization from the concentrate stream resulted in an 80–83% reduction in calcium content and simultaneously a 57–60% reduction in sulphate content. This should enable concentrate recirculation and high overall desalted water recovery using a modified EDR unit.
Turek, M., Bandura, B.	2007	Renewable energy by reverse electrodialysis	Desalination 205 (2007) 67–74	The objective of our work was to assess the usefulness of a	Poland	Polish Ministry of Science and Information	reverse electrodialysis.	WT	The price of low resistance ion-exchange membranes may be reduced a hundred

(*continued*)

Authors	Year	Title	Journal name, volume (issue)	Abstract	Country	Funding Agent	Technology used	USE (WT/WWT)	Recommendations
				previously developed electrodialyzer equipped with a 0.19-mm spacer in the RED process.		Technology, and SUT project.			times, which seems to be the desired cost level.
Kujawski, W., Krajewska, S., Kujawski, M., Gazagnes, L., Larbot, A., Persin, M.	2007	Pervaporation properties of fluoroalkylsilane (FAS) grafted ceramic membranes	Desalination 205 (2007) 75–86	The aim of this work was to prepare FAS modified pervaporation ceramic membranes by grafting an alumina and/or titania ceramic NF membrane with 1H, 1H, 2H, 2H-perfluorodecyl triethoxysilane (FAS) molecules and to determine the properties of such modified hydrophobic ceramic membranes in the pervaporation process applied for the desalination of different salt solutions and for separation of water–ethanol and water–MTBE mixtures.	Poland	Wojciech Kujawski at the Institut Européen des Membranes, CNRS (UMR5635) in Montpellier Socrates–Erasmus mobility grant, which enabled research training at the Ecole Nationale Superieure de Chimie de Montpellier and at the Institut Européen des Membranes CNRS-ENSCM-UM II (Montpellier, France).	Fluoroalkyl silanes grafted ceramic membranes.	WT	In the case of pervaporation of water–electrolyte solutions, only water was found in the permeate. The pervaporation flux was inversely proportional to the salt concentration in the feed. It was also found that FAS grafted membrane selectivity was dependent on the concentration of the grafting solution. In the case of more hydrophobic MTBE molecules, the tangled–straightened change of FAS molecules occurred in the concentration range 0–0.1 mol. % MTBE.
Turek, M., Dydo, P., Trojanowska, J., Bandura, B.	2007	Electrodialytic treatment of boron-containing wastewater	Desalination 205 (2007) 185–191	The objective of the present work was to investigate the technical and economic possibility of electrodialytic treatment of Tarnowskie Gory landfill leachate, described elsewhere.	Poland	These studies were supported by the following projects: Polish Minister of Science and Information Technology; NATO-CLG project and SUT project.	ED	WWT/ Industrial	The estimated total cost of leachate treatment in proposed two stage electrodialysis found was equal to $\$1.27/m^3$.

(continued)

Authors	Year	Title	Journal name, volume (issue)	Abstract	Country	Funding Agent	Technology used	USE (WT/WWT)	Recommendations
Turek, M., Dydo, P., Trojanowska, J., Campen, A.	2007	Adsorption/co-precipitation—reverse osmosis system for boron removal	Desalination 205 (2007) 192–199	Identify the suitable solid boron-rich concentrate obtainment method as well as, based on our previous results, propose an integrated system that matches the above requirement.	Poland	These studies were supported by the following projects: Polish Minister of Science and Information Technology, NATOCLG project and SUT project.	RO,ED.	WWT/ Industrial	It was also shown that the adsorption/co-precipitation step should be operated at a high boron content in order to avoid large precipitant load and resulting high RO feed-water salinity.
Abdul-Fattah, A.F.	1986	Supply of desalted water to remote arid zones	Desalination, 60 (1986) 151–164	Three options for supply of water from desalination plants to isolated population pockets are considered here.	Saudia Arabia	NA	Desalination	WT/SW	Provide an sample source of water appropriate for human consumption and to guarantee a continuous supply of water.
Bushnak, A.A.	1990	Water Supply Challenge in the Gulf Region	Desalination, 78 (1990) 133–145	This paper highlights some of the water challenges facing the Gulf region and recommends some key policies and programs to turn these challenges into opportunities.	Saudia Arabia	NA	Desalination	WT/SW	Adopt and implement consistent water policies leading to a regional water plan. Reduce water subsidies specially for major users. Increase water supply options under all conditions in both short and long terms. Create competition in water supply institutions. Reduce waste and increase utilization of available water sources.
AL-Malack, M.H., Anderson, G.K.	1996	Coagulation-crossflow microfiltration of domestic wastewater	Journal of Membrane Science 121 (1996) 59–70	The effect of using alum, polyaluminum silicate sulphate (PASS), and lime as coagulants on the performance of crossflow microfiltration of domestic wastewater was investigated.	Saudia Arabia	Research Institute of King Fahd University of Petroleum and Minerals (Dhahran, Saudi Arabia).	UF, MF, RO.	WWT/ Domestic	Lime was not found to be a suitable coagulant and this was attributed to the pH value used in the study.

(*continued*)

Authors	Year	Title	Journal name, volume (issue)	Abstract	Country	Funding Agent	Technology used	USE (WT/WWT)	Recommendations
Hassan, A.M., Farooque, A.M., Jamaluddin, A.T.M., AL-Amoudi, A.S., AL-Sofi, M.AK., AL-Rubaian, A.F., Kither, N.M., AL-Tisan, I.A.R., Rowaili, A.	2000	A demonstration plant based on the new NF-SWRO process	Desalination 131 (2000) 157–171	This paper describes the NF-SWRO demonstration plant, its operation and performance under different operating conditions utilizing different NF membranes in combination with SWRO membranes.	Saudia Arabia	NA	NF & RO.	WT/SW	The study allowed for the establishment of operating parameters for future NF-SWRO plants. These operating parameters shall be utilized in the operation of the now under construction NF-SWRO plant at Umm Lujj, Saudi Arabia.
Khedr, M.G.	2002	Development of reverse osmosis desalination membranes composition and configuration: future prospects	Desalination 153 (2002) 295–304	The present work reports on twelve-year experience in production, application, research and development related to RO membranes and systems by SIDh4AS Ltd. Particular emphasis is made on the retrofitting of RO systems originally designed to use CA or HF membranes to use the TFC, PA membranes. Encountered problems and realized advantages are discussed.	Saudia Arabia	NA	RO	WT/SW	The replacement of the RO flat CA and the HF membranes by TFC,PA ones positively contributed to the advance of the technology and the solution of the main problems which determine the RO process efficiency and cost-effectiveness.
Hamed, O.A., Mardouf, K.B., Al-Omran, A.	2007	Impact of interruption of antiscalant dosing or cleaning balls circulation during MSF plant operation	Desalination 208 (2007) 192–203	The aim of this study is to establish operating conditions and consequences when antiscalant dosing or ball circulation is interrupted.	Saudia Arabia	SWDRI and Al-Khobar Power & Desalination Plant.	MSF	WT/Sea water	It is recommended that plants which are using phosphonate based antiscalant have to review the corrective actions which they normally follow in case of interruption of ball circulation. Al-Khobar Phase-II plant evaluation test revealed that it is quite

(*continued*)

Authors	Year	Title	Journal name, volume (issue)	Abstract	Country	Funding Agent	Technology used	USE (WT/WWT)	Recommendations
									safe to operate the MSF distiller without antiscalant dosing for a maximum period of 3h.
Ng, H.Y. Elimelech, M.	2004	Influence of colloidal fouling on rejection of trace organic contaminants by reverse osmosis	Journal of Membrane Science 244 (2004) 215–226	A systematic investigation on the influence of colloidal fouling on removal of trace organics (steroid hormones) and inert organics by reverse osmosis (RO) membranes is reported.	Singapore	US National Science Foundation and the National University of Singapore Postdoctoral Training Fellowship to H.Y. Ng.	RO	WT/organic matter	Development of a colloidal cake layer on the membrane surface restricted back diffusion of both salt and inert organic solutes, resulting in significant decline in their rejection.
Sofia, A., Ng, W.J., Ong, S.L.	2004	Engineering design approaches for minimum fouling in submerged MBR	Desalination 160 (2004) 67–74	This paper compares the effectiveness of fine and coarse bubble diffusers in inducing crossflow velocity over a range of aeration intensities. Additionally, this paper considers the effectiveness of off-site chemical and mechanical cleaning.	Singapore	Kubota Co. (Japan).	MBR	WWT/ Domestic	Off-site chemical cleaning was effective in restoring subsequent membrane performance between cleanings. The air scouring effect was able to prolong the membrane operational lifespan by dislodging foulants from the membrane surface. A combined approach utilizing chemical cleaning and fine aeration bubbles could be an effective means in minimizing fouling and obtaining better as well as more stable performance for the submerged MBR.
Fane, A.G., Yeo, A., Law, A., Parameshwaran, K., Wicaksana, F., Chen, V.	2005	Low pressure membrane processes doing more with less energy	Desalination 185 (2005) 159–165	This paper discusses strategies to reduce energy usage in low pressure membrane processes. Studies on the role of air bubbling to control fouling in submerged membrane systems are described.	Singapore	ASTAR in Singapore, AusAid and the ARC in Australia.	MF&UF	WWT/ Industrial& Municipal	In low pressure membrane processing of concentrated feeds, such as the MBR, the use of bubbling to control fouling is important.

(continued)

Authors	Year	Title	Journal name, volume (issue)	Abstract	Country	Funding Agent	Technology used	USE (WT/WWT)	Recommendations
Qin, J.-J., Oo, M.H., Wai, M.N., Kekre, K.A.	2005	TOC removal in reclamation of municipal wastewater by RO	Separation and Purification Technology 46 (2005) 125–128	A pilot study on total organic carbon (TOC) removal in reclamation of municipal wastewater by RO has been conducted in Singapore. Secondary effluent from final clarifier was used as a feed to the system.	Singapore	Ms. Hsiaowan Lee and Mr. Wang Yuan Jun, Raymond for their contributions and the support from the Public Utilities Board of Singapore in the study.	RO	WWT/ Municipal	On-line TOC measurements of the RO permeate were in the range of 36–66 ppb at operating pressure of 60 psi for RE4040-FL membrane. TOC and salt rejections of the RE4040-FE membrane were higher than 99.5 and 97.8%, respectively. A high normalized flux of $30 Lm^{-2} h^{-1} MPa^{-1}$ at 25 °C was noted for RE4040-FE RO membrane at operating pressure of 70psi and water recovery of 50%.
Xi, W., Rong, W., Zhansheng, L.G., Fane, A.G.	2006	Development of a novel electrophoresis-UV grafting technique to modify PES UF membranes used for NOM removal	Journal of Membrane Science 273 (2006) 47–57	This study was an attempt to develop an effective method to modify UF membranes to meet the NOM removal requirement and to control the fouling caused by NOM in the filtration process.	Singapore	Agency of Science, Technology and Research of Singapore (A*STAR) for funding the Temasek Professor Program in Membrane Technology for Sustainable Water. The Public Utilities Board of Singapore is thanked for supplying the raw water samples.	UF	Raw water (NOM)	It was found that the modified membrane surfaces exhibited more hydrophilic and negative charged features after the electrophoresis-UV grafting treatment. The MA and AAG grafting reduced the MWCO of the PES membranes to improve NOM retention, and decreased the permeability to less than half of the original.
Li, Z., Ren, J.R., Fane, A.G., Li, D.F., Wong, F.S.	2006	Influence of solvent on the structure and performance of cellulose acetate membranes	Journal of Membrane Science 279 (2006) 601–607	In this study, cellulose acetate was chosen as the polymer due to its extensive use as membrane material covering from gas separation to microfiltration. Furthermore, the inherent fouling resistant property due	Singapore	This work was supported by the Agency for Science, Technology and Research, Singapore, Temasek Professor Program on Membrane Technology for Sustainable Water.	CA	CA/ NMP/ H_2O	The membrane structure was tunable by changing the ratio of NMP/GBL. A critical ratio of NMP/GBL appears to exist that denotes the point of change of membrane structure from macrovoids to sponge-like.

(continued)

Authors	Year	Title	Journal name, volume (issue)	Abstract	Country	Funding Agent	Technology used	USE (WT/WWT)	Recommendations
				to its neutral surface and hydrophilicity make cellulose acetate a popular membrane material for water and wastewater treatment.					
Chin, S.S., Chiang, K., Fane, A.G.	2006	The stability of polymeric membranes in a TiO_2 photocatalysis process	Journal of Membrane Science 275 (2006) 202–211	This study focuses on the selection of polymeric membranes for the photocatalytic-membrane process. Ten membranes were evaluated under ultraviolet (UV) and TiO_2 photocatalysis conditions.	Singapore	ASTAR (Singapore) as part of the Temasek Professorship Programme on Membrane Technology for Sustainable Water. Dr. Augustine Tan from National Institute of Education (NIE), NTU is thanked for help in FTIR analysis.	Polymeric membrane.	WT	The current work has successfully provided a protocol for testing the stability of polymeric membranes before they can be used with photocatalytic reactions.
Yang, P., Teo, W.-K., Ting, Y.-P.	2006	Design and performance study of a novel immobilized hollow fibber membrane bioreactor	Bio resource Technology 97 (2006) 39–46	The purpose of this study is to develop an immobilized cell bioreactor using these hollow fibber membranes.	Singapore	NA	hollow fibber membrane bioreactor.	nutrients	The dual-layer hollow fibber bioreactor investigated in this study was effective in eliminating the mass transfer limitation of liquid substrate in the culture of E. coli.
Chin, S.S., Lim, T.M., Chiang, K., Fane, A.G.	2007	Hybrid low-pressure submerged membrane photoreactor for the removal of bisphenol A	Desalination 202 (2007) 253–261	The objective of this study was to investigate the efficiency of a low-pressure SMPR in the degradation of an endocrine disrupting chemical using commercial TiO2, Degussa P25.	Singapore	ASTAR (Singapore)	SMPR	WT/Industrial	The compact and low-pressure SMPR has shown to be effective and attractive for the photodegradation of low concentrations of organics in water.
Fane, A.G.	2007	Sustainability and membrane processing of wastewater for reuse	Desalination 202 (2007) 53–58	levels of sustainability in the context of membrane technology and water reclamation.	Singapore	UNESCO Centre for Membrane Science & Technology, UNSW, in Sydney and in the Temasek Professor	RO	WWT/ Municipal	The major limitation is membrane fouling and that influences the selection of conditions for long-term sustainable operation.

(continued)

Authors	Year	Title	Journal name, volume (issue)	Abstract	Country	Funding Agent	Technology used	USE (WT/WWT)	Recommendations
						Programme, NTU, in Singapore. A*STAR (Singapore).			
Lee, J., Ahn, W.-Y., Lee, C.-H.	2001	Comparison of the filtration characteristics between attached and suspended growth microorganisms in submerged membrane	Wat. Res. Vol. 35, No. 10, pp. 2435–2445, 2001	An attached growth bioreactor was designed to minimize the effect of suspended microorganisms on membrane fouling in submerged membrane bioreactor.	South Korea	Korean Ministry of Science & Technology	SMBR	suspended growth.	The loss of membrane permeability proceeded more rapidly with the attached growth system than with the suspended one. Better filtration performance with suspended growth was attributed to the role of dynamic membrane formed on the membrane surface, which was confirmed by SEM and AFM images. Better filterability with the suspended growth could also be due to the rougher cake layer having smaller specific cake resistance than that with the attached growth.
Ahn, W.-Y., Kang, M.-S., Yim, S.-K., Choi, K.-H.	2002	Advanced landfill leachate treatment using an integrated membrane process	Desalination 149 (2002) 109–114	A landfill leachate treatment plant (LLTP) in Korea was retrofitted to improve treatment efficiency by using integrated membrane process that is composed of membrane bioreactor (MBR) and reverse osmosis (RO) process.	South Korea	NA	MBR&RO.	Landfill leachate.	The integrated membrane process greatly enhances treatment efficiency and economically removes none biodegradable organic and nitrogenous matters of landfill leachate.
Leeavb, Y., Cho, J., Sea, Y., Lee, J.W., Ahn, K.-H.	2002	Modeling of submerged membrane bioreactor process	Desalination 146 (2002) 451–457	In this study, a mathematical model has been developed for the submerged	South Korea	Green Korea 21 project founded by the Korea Institute of Science and	SMBR	WWT/ Municipal	The model presented here shall be useful in optimizing operating parameters in the aspects of effluent quality

(*continued*)

Authors	Year	Title	Journal name, volume (issue)	Abstract	Country	Funding Agent	Technology used	USE (WT/WWT)	Recommendations
		for wastewater treatment		membrane bioreactor (SMBR) combining the activated sludge model No.1 (ASMI) with a membrane fouling model.		Technology (KIST), Korea.			as well as membrane fouling control.
Shim, J.K., Yoo, I.K, Lee, Y.M.	2002	Design and operation considerations for wastewater treatment using a flat submerged membrane bioreactor	Process Biochemistry 38 (2002) 279–285	The objective of this study was to investigate the effect of some design and operational variables on the proper control of membrane clogging in MBR wastewater treatment process.	South Korea	University of Ulsan Research Fund of 2002	MBR	WWT/ Municipal	Due to the distinctive advantages of the membrane bioreactor, the COD removal efficiency and the nitrification performance was satisfactory for a high concentration of organics and nitrogen compounds.
Bae, T.-H., Han, S.-S., Tak, T.-M.	2003	Membrane sequencing batch reactor system for the treatment of dairy industry wastewater	Process Biochemistry 39 (2003) 221–231	A membrane separation process was coupled to a sequencing batch reactor (SBR) for biological nutrient removal (BNR) processes and a combined system was named a membrane sequencing batch reactor (MSBR).	South Korea	NA	MSBR	WWT/ Industrial	The high phosphorus concentration of the influent gave low removal efficiency due to the limitation of the biological removal process. Nitrifying bacteria was not cultivated adequately due to the high BOD: TKN ratio of the influent. A diffuser-attached module design, subcritical flux operation, BOD removal of the MSBR system was high (97/98%) and stable regardless of operation modes.
Lee, S. Cho, J., Elimelech, M.	2004	Influence of colloidal fouling and feed water recovery on salt rejection of RO and NF membranes	Desalination 160 (2004) 1–12	The objective of this study was to systematically investigate the influence of colloidal fouling on the salt rejection behaviour of RO and NF membranes.	South Korea	This research was supported by the US National Science Foundation (Grant BES 0114527), the Brain Korea 21 (BK21) Project from the Ministry of Education in Korea, and the Korean Science and	RO, NF	WT/ Salt water	The NF membrane suffered a much greater decline in salt rejection than RO, most likely due to a substantial decrease in charge (Donnan) exclusion as salt concentration at the membrane surface increases.

(*continued*)

Authors	Year	Title	Journal name, volume (issue)	Abstract	Country	Funding Agent	Technology used	USE (WT/WWT)	Recommendations
						Engineering Foundation (KOSEF) through the Advanced Environmental Monitoring Research Center (ADEMRC) at K-JIST.			
Chang, In.-S., Kim, S.-N.	2005	Wastewater treatment using membrane filtration—effect of biosolids concentration on cake resistance	Process Biochemistry 40 (2005) 1307–1314	The effect of biosolids concentrations on permeate flux during the membrane filtration of activated sludge suspensions with different mixed liquor of suspended solids (MLSS) concentration was investigated.	South Korea	NA	UF	WWT/ Synthetic	From the point of view of membrane fouling, tertiary wastewater treatment with membrane does not have any beneficial effects compared to MBR.
Yeon, K.-M, Park, J.-S., Lee, C.-H., Kim, S.-M.	2005	Membrane coupled high-performance compact reactor: A new MBR system for advanced wastewater treatment	Water Research 39 (2005) 1954–1961	This study examined the potential and limitations of a new submerged membrane system coupled with a High performance Compact Reactor (HCR) to take advantages of both systems.	South Korea	Korean Ministry of Science & Technology	MHCR,MBR	synthetic waste-water	In MHCR, an open-type configuration of membrane module resulted in less membrane fouling than a fixed-type module due to its higher mobility. The optimum location of a membrane module is at the bottom, inside the draft tube. The MHCR showed much greater membrane permeability than the conventional MBR, promising very high potential for the treatment of wastewater of high organic strength.
Yun, M.-A., Yeon, K.-M., Park, J.-S., Lee, C.-H., Chun, J., Lim, D.J.	2006	Characterization of biofilm structure and its effect on membrane	Water Research 40 (2006) 45–52	The aim of this study was to elucidate the effect of dissolved oxygen on membrane	South Korea	Korean Ministry of Science & Technology	MBR	Dye waste-water	Membrane filterability was closely associated with the structural parameters. In the aerobic MBR, the

(*continued*)

Authors	Year	Title	Journal name, volume (issue)	Abstract	Country	Funding Agent	Technology used	USE (WT/WWT)	Recommendations
		permeability in MBR for dye wastewater treatment		biofouling in terms of the structural analysis of biofilms.					accumulation of biomass with the growth of biofilm was the main factor for the loss of permeability, faster biofouling was observed in anoxic MBR compared to the aerobic one.
Jin, Y.-L., Lee, W.-H., Lee C. -H., Chang, I.-S., Huang, X., Swaminathan, T.	2006	Effect of DO concentration on biofilm structure and membrane filterability in submerged membrane bioreactor	Water Research 40 (2006) 2829–2836	The structures of biofilms deposited on the membrane surface under different dissolved oxygen (DO) conditions were characterized to identify its relation to membrane filterability in membrane bioreactors (MBR).	South Korea	Korean Ministry of Science & Technology.	MBR	WWT/ Municipal	The rate of membrane fouling for the LDO reactor was 7.5 times faster than that of the HDO reactor. Differences in biofilm characteristics as a result of different DO level were the main factors affecting biofouling for both MBRs.
Jung, Y.-J., Koh, H.-W., Shin, W.-T., Sung, N.-C.	2006	A novel approach to an advanced tertiary wastewater treatment: Combination of a membrane bioreactor and an oyster-zeolite column	Desalination 190 (2006) 243–255	A combination of a microfiltration-membrane bioreactor (MBR) and oyster-zeolite (OZ) packed-bed adsorption column was studied for the first time to evaluate the advanced tertiary treatment of nitrogen and phosphorous.	South Korea	Oak Ridge National Laboratory.	MBR & oyster-zeolite column.	WWT/ Municipal	The results of this study demonstrate a possible method of oyster shell waste utilization in South Korea.
Kim, In. S., Jang, N.	2006	The effect of calcium on the membrane biofouling in the membrane bioreactor (MBR)	Water Research 40 (2006) 2756–2764	A continuous bench-scale submerged membrane bioreactor (BSMBR) was operated in two modes; low (LC) and optimum calcium (OC) concentrations,	South Korea	Sustainable Water Resources Research Center (SWRRC) of the 21st century frontier R&D program and, in part, by the Center for Distributed Sensor Network at Gwangju	BSMBR	WWT/ Synthetic.	A continuous bench-scale submerged membrane bioreactor (BSMBR) was operated in two modes; low (LC) and optimum calcium (OC) concentrations, to investigate the effect of calcium on membrane

(*continued*)

Authors	Year	Title	Journal name, volume (issue)	Abstract	Country	Funding Agent	Technology used	USE (WT/WWT)	Recommendations
				to investigate the effect of calcium on membrane biofouling.		Institute of Science and Technology (GIST).			biofouling. The FTIR analysis of the fouled membrane surface supported that proteins and carbohydrates were main foulants during the BSMBR operation.
Kim, Y., Giokas, D.L., Lee, J.-W., Paraskevas, P.A.	2006	Potential of natural treatment systems for the reclamation of domestic sewage in irrigated agriculture	Desalination 189 (2006) 229–242	The present study is focused on the investigation of several configurations of natural treatment systems as means for cost-effective treatment of municipal sewage.	South Korea	Basic Research Program of the Korea Science and Engineering Foundation.	natural treatment systems.	WWT/ Municipal	The results indicate that this system may be favourable for various wastewater reuse applications based on existing reuse criteria or guidelines.
Ma, B.-C., Lee, Y.-N., Park, J.-S., Lee, C.-H., Lee, S.-H., Chang, I.-S., Ahn, T.-S.	2006	Correlation between dissolved oxygen concentration, microbial community and membrane permeability in a membrane bioreactor	Process Biochemistry 41(5) (2006) 1165–1172	The effect of dissolved oxygen (DO) level on microfiltration performance and microbial physiology was investigated in membrane-coupled sequencing batch reactors (MSBR).	South Korea	Korean Ministry of Science and Technology.	MSBR	WWT/ Municipal	TMP value due to the head loss originated from the water level reduction was calculated to 2.0 kPa. TMP build-up was more dependent on flux and airflow rate in the A/O than in the –O–. Worse filtration performance in the A/O was attributed to smaller size of particles and thus larger specific cake resistance. Irreversible membrane fouling in the A/O was more important than in the –O– so that MSBR with the A/O phase resulted in faster TMP build-up. FISH and DGGE analyses revealed that the microbial community structures in two reactors were different.

(*continued*)

Authors	Year	Title	Journal name, volume (issue)	Abstract	Country	Funding Agent	Technology used	USE (WT/WWT)	Recommendations
Chae, S.R., Shin, H.S.	2006	Effect of condensate of food waste (CFW) on nutrient removal and behaviours of intercellular materials in a vertical submerged membrane bioreactor (VSMBR)	Bioresource Technology 98(2) (2007) 373–379	The main objective of this study was to investigate the effect of condensate of food waste (CFW) on nutrient removal in a pilot-scale vertical submerged membrane bioreactor (VSMBR) treating municipal wastewater having total-chemical oxygen demand to total-nitrogen ratio (T-COD/T-N) of 5.5.	South Korea	Ministry of Environment as 'The Eco-technopia 21 project'.	VSMBR	Condensate of food waste	VSMBR with the CFW was regarded as an effective and compact process to treat low strength wastewater as it showed high removal efficiencies of not only COD but also nutrients in the system.
Kim, K.S., Kwon, T.S., Yang, J.S., Yang, J.W.	2007	Simultaneous removal of chlorinated contaminants by pervaporation for the reuse of a surfactant	Desalination 205 (2007) 87–96	In this study, the simultaneous removal of TCE and PCE from surfactant solution containing alcohols was investigated.	South Korea	National Research Laboratory Program of the Korean Ministry of Science and Technology.	Pervaporation	chlorinated solvents.	Organic flux was significantly influenced by flow rate, feed concentration and temperature The important parameter for selectivity was temperature because all flux, especially water flux, increased remarkably due to activated molecular mobility and membrane swelling.
Park, J.S., Song, J.H., Yeon, K.H., Moon, S.H.	2007	Removal of hardness ions from tap water using electromembrane processes	Desalination 202 (2007) 1–8	In this study, the performances of EDR and EDIR were investigated for removal of hardness ions in tap water, and the performances for the removal of hardness materials were compared with an ED process.	South Korea	This work was supported by the grant (4–1-2) from the Sustainable Water Resource Research Center of the 21st Century Frontier R&D Program through the Water Reuse Technology Center at Gwangju Institute of Science and Technology.	ED	WT/Tap water.	This study showed the performances of EDR and EDIR systems as an alternative process of ion exchange for removal of hardness ions in tap water.

(continued)

Authors	Year	Title	Journal name, volume (issue)	Abstract	Country	Funding Agent	Technology used	USE (WT/WWT)	Recommendations
Lee, C.W., Bae, S.D., Han, S.W., Kang, L.S.	2007	Application of ultrafiltration hybrid membrane processes for reuse of secondary effluent	Desalination 202 (2007) 239–246	This study evaluates the factors affecting pretreatment conditions for hybrid UF membrane processes for reuse of secondary effluent from the sewage treatment plant.	South Korea	NA	UF	WWT/ Municipal	Hybrid UF membrane processes can be applied for direct reuse of secondary effluent as a type of urban reuse, agricultural reuse, landscape and industrial reuse or as a pretreatment before a second step of a reverse osmosis membrane process.
Kang, Y.-T., Cho, Y.-H., Chung, E.-H.	2007	Assessing the feasibility of a vertical double submerged membrane modules system for wastewater reclamation under different conditions.	Desalination 202 (2007) 59–67	This paper is intended to help move towards optimal design on operation of submerged reactor treatment plants. The optimum composition of a new reclamation system, satisfying a more stringent regulation of reclamation in order to reuse municipal wastewater has been studied.	South Korea	NA	Submerged membrane bioreactor (SMBR).	WWT/ Municipal	As the results the vertical double membrane modules configuration can produce twice the quantity of water with the same bioreactor volume of the simple membrane module.
Jang, N., Ren, X., Kim, G., Ahn, C., Cho, J., Kim, I.S.	2007	Characteristics of soluble microbial products and extracellular polymeric substances in the membrane bioreactor for water reuse.	Desalination 202 (2007) 90–98	In this study, the characteristic of SMP and EPS were investigated using a submerged membrane bioreactor, which is one of the most promising water reuse process.	South Korea	Sustainable Water Resources Research Center (SWRRC) of the 21st century frontier R&D program and the Center for Distributed Sensor Network at the Gwangju Institute of Science and Technology (GIST).	membrane bioreactor.	WWT/ Municipal	Both the soluble and SS parts of the MFI increased with increasing F/M ratios. The soluble MFI increased with increases in the rejection of the COD concentrations. It was assumed that the SS MFI increased due to the higher EPS concentrations. The MFI during denitrification showed higher values than those during nitrification, possibly due to floc deterioration.

(*continued*)

Authors	Year	Title	Journal name, volume (issue)	Abstract	Country	Funding Agent	Technology used	USE (WT/WWT)	Recommendations
Kang, Y.-T., Cho, Y.-H., Chung, E.-H.	2007	Development of the wastewater reclamation and reusing system with a submerged membrane bioreactor combined bio-filtration	Desalination 202 (2007) 68–76	This paper is intended to help move towards an optimal design on the operation of submerged reactor treatment plants. The optimum composition of a new reclamation system satisfying a more stringent regulation of reclamation in order to reuse municipal wastewater has been studied.	South Korea	NA	Bio-filter membrane bioreactor (BMB) system	WWT/ Domestic	Not only does the study for improving the colour removal, combined with the pre-treatment or post-treatment need to be furthered, but also the economical analysis is needs to be investigated.
Kim, J.-O., Jung, J.-T., Yeom, I.-T., Aoh, G.-H.	2007	Effect of fouling reduction by ozone backwashing in a microfiltration system with advanced new membrane material	Desalination 202 (2007) 361–368	The purpose of this research is to investigate the effect of ozone backwashing for permeation flux recovery.	South Korea	Regional Technology Innovation Program of the Ministry of Commerce, Industry and Energy and partially supported by the Eco-STAR project of the Ministry of Environment.	MF	WWT/ Industrial	It is favourable to operate membrane cleaning before the foulant is consolidated on the membrane surface.
Lim, B.S., Choi, B.C., Yu, S. W., Lee, C.G.	2007	Effects of operational parameters on aeration on/off time in an intermittent aeration membrane bioreactor	Desalination 202 (2007) 77–82	This study was carried out to investigate the optimum working conditions and to evaluate the effects of operational parameters in an intermittent aeration membrane bioreactor.	South Korea	Basic Research Program of the Korea Science & Engineering Foundation.	aeration membrane bioreactor	WWT/ Domestic	Microbial activity was affected only a little by variation in aeration on/off time, BOD removal efficiency rate of this process was higher than 97% regardless of the aeration on/off time distribution and reached a maximum of 99% at 40/80 min aeration on/off time. EPS per unit of microorganisms increased as the aeration off time got longer.

(*continued*)

Authors	Year	Title	Journal name, volume (issue)	Abstract	Country	Funding Agent	Technology used	USE (WT/WWT)	Recommendations
Kim, J.-O., Jung, J.-T., Yeom, I.-T., Aoh, G.-H.	2007	Electric fields treatment for the reduction of membrane fouling, the inactivation of bacteria and the enhancement of particle coagulation	Desalination 202 (2007) 31–37	The objective of this study is to evaluate the effect of electric fields application for the reduction of membrane fouling, the inactivation of microorganisms and the enhancement of particle coagulation.	South Korea	Korea Research Foundation and partially supported by the Eco-STAR project of the Ministry of Environment.	MF	WWT/ Municipal	This non-thermal electrical fields process is anticipated as energy saving and economical technology and is expected to improve the efficiency of the process by providing electric fields as pre treatment step.
Park, N., Lee, S., Yoon, S.R., Kim, Y.H., Cho, J.	2007	Foulants analyses for NF membranes with different feed waters: coagulation/ sedimentation and sand filtration treated waters	Desalination 202 (2007) 231–238	The objective of this study is to compare two different nanofiltration membranes with different pore size distribution, surface charge, and hydrophobicity, with respect to organic (NOM and microorganisms related substances) fouling.	South Korea	Korea Science and Engineering Foundation (KOSEF), Sustainability Water Resources Research Center of 21st Century Frontier Research Program.	NF	WT/Lake water	Inorganic foulants (i.e., Si and Al) were also fouled onto the membrane surface and/or pores and it is effectively removed by caustic cleaning, not pure water.
Kim, H.S., Seo, I. S., Kim, Y.K., Kim, J.Y., Ahn, H. W., Kim, I.S.	2007	Full-scale study on dynamic state membrane bio-reactor with modified intermittent aeration	Desalination 202 (2007) 99–105	The aim of this study is to evaluate the performance of MIA in the full-scale plant developed for applying to a dynamic state MBR as a BNR process.	South Korea	NA	MBR	WWT/ Municipal	Applied the membrane was continuously operated without out-line chemical cleaning by only in-line air scrubbing for 200 days.
Yeon, K.-H., Song, J.-H., Shim, J., Moon, S.-H., Jeong, Y.-U., Joo, H.-Y.	2007	Integrating electrochemical processes with electrodialysis reversal and electro-oxidation to	Desalination 202 (2007) 400–410	In the present study, the integration of the electrochemical processes with EDR and ANT is proposed as a treatment	South Korea	Korea Power Engineering Co., Inc. (KOPEC).	EDR	WWT/ Municipal	The ANT process, which has been proposed as a novel treatment strategy for EDR concentrates, has been successfully implemented in this study

(*continued*)

Authors	Year	Title	Journal name, volume (issue)	Abstract	Country	Funding Agent	Technology used	USE (WT/WWT)	Recommendations
		minimize COD and T-N at wastewater treatment facilities of power plants.		strategy to simultaneously remove the COD and T-N in the nuclear power plant of South Korea.					for the removal of hard to treat wastes such as EDR concentrates.
Kim, S., Hoek, E.M.V.	2007	Interactions controlling biopolymer fouling of reverse osmosis membranes	Desalination 202 (2007) 333–342	The theoretical approach used to understand organic fouling of RO membranes employs the interfacial force balance approach of Wang et al., which combines interfacial hydrodynamic and XDLVO interactions to describe colloidal deposition.	South Korea	Korea Science and Engineering Foundation (KOSEF), California Department of Water Resources through the Desalination Research Innovation Partnership (DRIP).	RO	WWT/ Municipal	In future studies, fouling of RO and NF membranes will be studied using more complex suspensions comprising bacteria, biopolymers, geopolymers, and inorganic colloids.
Kim, R.-H., Lee, S., Jeong, J., Lee, J.-H., Kim, Y.-K.	2007	Reuse of greywater and rainwater using fibber filter media and metal membrane	Desalination 202 (2007) 326–332	In this study, novel treatment options including lignocelluloses filter media and metal membranes were examined to reuse greywater and rainwater in office buildings.	South Korea	Sustainable Water Resources Research Center (SWRRC) of 21st century frontier R&D program, Korea Institute of Construction Technology (KICT).	fiber filter media and metal membrane.	WWT/ Greywater and rainwater	For the mixture of greywater and rainwater, the initial permeability of 0.5mm metal membrane was the best but it rapidly decreased with time.
Ahn, Y.T., Kang, S.T., Chae, S.R., Lee, C.Y., Bae, B.U., Shin, H.S.	2007	Simultaneous high-strength organic and nitrogen removal with combined anaerobic upflow bed filter and aerobic membrane bioreactor	Desalination 202 (2007) 114–121	Biological treatment of high-strength nitrogen wastewater by a combined anaerobic (upflow sludge bed filter, UBF)/aerobic (membrane bioreactor, MBR) system has been studied.	South Korea	Ministry of Environment as 'The Eco-technopia 21 project', NRL Program.	upflow bed filter and aerobic membrane bioreactor.	WWT/ Synthetic	Higher organic and total nitrogen removal (99 and 46%) was possible at various organic and nitrogen loading rates. Decreased alkalinity from 4,000 to 3,000mg/L as ($CaCO_3$) affected organic removal efficiency and denitrification in anaerobic UBF reactor. Rapid TMP increase was observed at the combined process.

(continued)

Authors	Year	Title	Journal name, volume (issue)	Abstract	Country	Funding Agent	Technology used	USE (WT/WWT)	Recommendations
Kim, J.-O., Jung, J.-T., Chung, J.	2007	Treatment performance of metal membrane microfiltration and electrodialysis integrated system for wastewater reclamation	Desalination 202 (2007) 343–350	The integrated system with microfiltration/ electrodialysis was proposed for wastewater reclamation.	South Korea	21st Frontier research project sustainable water resources research centre and partially supported by the Regional Technology Innovation Program.	MF&ED	WWT/ Municipal	Suggested wastewater reclamation system showed stable water qualities during operational period over 6 months. Therefore, it is expected to be an applicable and useful system for the reclamation of wastewater.
Kim, Y.-H., Hwang, E.-D., Shin, W.S., Choi, J.-H., Had, T.W., Choi, S.J.	2007	Treatments of stainless steel wastewater containing a high concentration of nitrate using reverse osmosis and nanomembranes	Desalination 202 (2007) 286–292	A nanomembrane (NTR 729HF) and three different types of reverse osmosis (RO) membranes (CPA2, LFC1, and ESPA1) were tested for the treatment of high concentration of nitrate wastewater from stainless steel industry.	South Korea	NA	RO&NF	Stainless steel WW(Industrial)	Result implicates that membrane separation could be an alternative for the treatment of nitrate from stainless steel wastewater.
Schippers, J.C., Verduuwv, J., Hofman, J.M.	1980	Pretreatment of reverse osmosis feedwater by means of in line coagulation	Desalination, 32 (1980) 103–112	The possibilities of removing salts from Rhine water by means of reverse osmosis depend very considerably on the efficiency of the usable pretreatment techniques.	The Netherlands	NA	RO	WT/Rhine river	It has been shown that these humic substances can be conclusively removed by adjusting the addition of ferric salts in the preceding coagulation process to a sufficiently high level.
Schippers, J.C., Hanemaayerh, J.H., Smolderss, C.A., Kostensek, A.	1981	Predicting flux decline of reverse osmosis membranes	Desalination.38 (1981) 339–348	A mathematical model predicting flux decline of reverse osmosis membranes due to colloidal fouling has been verified.	The Netherlands	The research in this paper was commissioned by the Netherlands Waterworks' Association and has partly been conducted within the framework of the cooperation	RO	WT/Surface water	The result of this research was, that the measured and calculated values of flux Gecline of Lne ultrafiltration membranes as a function of time agree reasonably well. The difference between the measured and calculated

(*continued*)

Authors	Year	Title	Journal name, volume (issue)	Abstract	Country	Funding Agent	Technology used	USE (WT/WWT)	Recommendations
						between the Provincial Waterworks of North-Holland, Wafilin B.V. and the Netherlands Waterworks' Testing and Research Institute KIWA N.V.			values may be explained by the assumption, that initially blocking filtration occurs instead of cake or gel filtration.
Boerlage, S.F.E., Kennedy, M.D., Bonne, P.A.C., Galjaard, G., Schippers, J.C.	1997	Prediction of flux decline in membrane systems due to particulate fouling	Desalination 113 (1997) 231–233	This research reports on developments of the Modified Fouling Index using ultrafiltration modules as a water quality indicator and to predict flux decline occurring in membrane filtration installations.	The Netherlands	NA	RO, NF & Uf	WT/DW	The MFI-UF was found to be a promising tool for measuring the colloidal fouling potential for reverse osmosis, nana- and ultrafiltration systems.
Oosterom, H., Galjaard, G., Nederlof, M.M., Schippers, J.C.	1998	Feasibility of micro- and ultrafiltration for the direct treatment of surface water: results of various pilot studies	Desalination 119 (1998) 275–276	Goals were to determine the effect of different water qualities on membrane performance and to determine the amount of membrane fouling while treating raw surface water. Similarities and differences in membrane performance and raw water quality were also sorted out.	The Netherlands	NA	MF&UF	WT/Surface water	Further research will be necessary to optimize the performance of the membranes to optimal operating conditions on the direct treatment of surface water.
Kennedy, M., Kim, S.M., Mutenyo, I., Broens, L., Schippers, J.	1998	Intermittent crossflushing of hollow fibber ultrafiltration systems	Desalination 118 (1998) 175–188	In this study, backwashing conditions such as pressure and duration were studied in order	The Netherlands	NA	UF	Backwash water	In this study, flux restoration was significantly improved by crossflushing directly before backwashing, as the

(continued)

Authors	Year	Title	Journal name, volume (issue)	Abstract	Country	Funding Agent	Technology used	USE (WT/WWT)	Recommendations
				to maximize the net flux per cycle.					accumulated cake was rapidly removed by the high crossflush velocity in the fibber.
Boerlage, S.F.E., Kennedy, M.D. Aniye, M.P., Abogrean, E.M., Galjaard, G., Schippers, J.C.	1998	Monitoring particulate fouling in membrane systems	Desalination 118 (1998) 131–142	This research aims at further exploring the MFI-UF test for monitoring particulate fouling. The above criteria will be investigated, using a selected polyacrylonitrile UF module from previous research for varying types of feed water.	The Netherlands	Novem (Netherlands Agency for Energy and Environment) within the Subsidy Programme of New Technology for Industry (MINT).	MFI-UF	WT	The selected polyacrylonitrile module gave a stable and reliable value over time for high and low fouling feedwaters. For more fouling waters the MFI-UF test can be carried out at lower pressures and corrected to the reference pressure of the MFI-UF test.
Galjaard, G., Schippers, J.C., Nederlof, M.M., Oosterom, H.A.	1998	Quick-Scan: Selection of micro- and ultrafiltration membranes	Desalination 117 (1998) 79–84	The Quick-Scan (QS) is a quick, reproducible and relative cheap measuring method to compare micro- or ultrafiltration membrane performances within standard conditions.	The Netherlands	NA	MF&UF	WT	The Quick-Scan method is also a useful tool for a location where a full-scale installation is already implemented. The Quick-Scan can then be applied for research on new membranes or new operation conditions.
Bremere, I., Kennedy, M., Michel, P., Emmerik, R.V., Witkamp, G.J., Schippers, J.	1999	Controlling scaling in membrane filtration systems using a desupersaturation unit	Desalination 124 (1999) 51–62	This paper focuses on the effect of adsorption of organic matter and ozone on the precipitation of barium sulphate in a laboratory scale desupersaturation unit.	The Netherlands	NOVEM (Netherlands Agency for Energy and Environment) within the Subsidy Programme of New Technology for Industry (MINT), and commercial partners: Advanced Treatment Processes Europe (ATP), Amsterdam Water Supply (AWS), Heineken, Hydranautics Europe,	RO, NF, Desuper saturation	WT/DW	The effect of antiscalant and (natural) organic matter on barium sulphate growth retardation in AWS RO were similar. Irreversible inactivation of barium sulphate seed crystals (grain size 2.0–3.15mm) occurred in the AWS RO (without antiscalant), while the seed crystals were restored completely in the WSO NF,

(*continued*)

Authors	Year	Title	Journal name, volume (issue)	Abstract	Country	Funding Agent	Technology used	USE (WT/WWT)	Recommendations
						Kiwa N.V. Research and Consultancy and Overijssel Water Supply (OWS).			The pretreatment of feedwater had a strong effect on barium removal in RO and NF concentrates. The composition of organic matter was altered by ozonation.
Boerlage, S.F.E., Kennedy, M.D., Aniye, M.P. Abogrean, E.M., EL-Hodali, D.E.Y., Tarawneh, Z.S., Schippers, J.C.	2000	Modified Fouling Indexultrafiltration to compare pretreatment processes of reverse osmosis feedwater	Desalination 131 (2000) 201–214	This research investigates the application of the newly developed MFI-UF test as a tool to measure the particulate fouling potential of RO/NF feedwater.	The Netherlands	Kiwa N.V. Research and Consultancy and Novem (Netherlands Agency for Energy and Environment). J. Kruithof, H. Fohner and H. Scheerman at the PWN Water Supply Company of North Holland and J.P. van der Hoek and P. Bonn6 at Amsterdam Water Supply.	MFI-UF & RO	WT	In conclusion, the MFI-UF was found to be a promising tool for measuring the particulate fouling potential of a feedwater. Moreover, it can be used alone or in combination with the MFI0.45 or MFI0.05 to compare the efficiency of various pretreatment processes for the removal of selected particle sizes. Future work focuses on predicting flux decline in the MFI-UF test for membrane filtration systems.
Bremere, I., Kennedy, M., Stikker, A., Schippers, J.	2001	How water scarcity will effect the growth in the desalination market in the coming 25 years	Desalination 138 (2001) 7–15	This study focuses on quantifying the expansion of the desalination market with respect to providing an alternative water supply to municipalities and prevent the worsening of water scarcity due to projected population growth in various regions of the world.	The Netherlands	NA	Desalination	WT/Sea water	A growth in desalination capacity of ca. 15 million m^3/d will be needed to supply the urban population in the 10 water scarce countries and to prevent groundwater withdrawals for municipal purposes in the next 25 years.

(*continued*)

Authors	Year	Title	Journal name, volume (issue)	Abstract	Country	Funding Agent	Technology used	USE (WT/WWT)	Recommendations
Kennedy, M.D., Chun, H.K., Yangal, V.A.Q., Heijman, B.G.J., Schippers, J.C.	2005	Natural organic matter (NOM) fouling of ultrafiltration membranes: fractionation of NOM in surface water and characterisation by LC-OCD	Desalination 178 (2005) 73–83	The aim of this study was to obtain a better understanding of the interactions between the fractional components of NOM and a hydrophilic PES/ PVP hollow fibber ultrafiltration membrane (150–200kDa).	The Netherlands	NA	UF	NOM&Surface Water	The use of XAD-8/4 resins to fractionate NOM gave a lot of overlap betweens hydrophobic and hydrophilic compounds. The fouling rates were in the order hydrophilic NOM > transphilic NOM > hydrophobic NOM. The DOC rejection of the UF membrane was <1%.
Evenblij H., Geilvoet, S., van der Graaf, J.H.J.M., van der Roest, H.F.	2005	Filtration characterisation for assessing MBR performance: three cases compared	Desalination 178 (2005) 115–124	This article describes the results of applying the mentioned filtration characterisation method at three MBR pilot locations in the Netherlands.	The Netherlands	DHV Water.	MBR	WWT/ Municipal	Further research into the substances involved in fouling will eventually lead to focusing on the exact (group of) substances that cause membrane fouling. Once this is accomplished, the most important step is still to be taken: cultivation of the biomass in order to minimise the release of these substances to the water phase.
Nederlof, M.M., van Paassen, J.A.M., Jong, R.	2005	Nanofiltration concentrate disposal: experiences in The Netherlands	Desalination 178 (2005) 303–312	Several ground water sources show elevated concentrations of hardness, colour, sulphate and/ or organic micropollutants.	The Netherlands	NA	NF	WT/Surface water	For the future, additional techniques should be developed for removing critical components from the concentrate or to prevent concentrate discharge at all (by using crystallization and/or evaporation).
Meindersma, G.W., Guijt, C.M., de Haan, A.B.	2006	Desalination and water recycling by air gap membrane distillation	Desalination 187 (2006) 291–301	Membrane distillation differs from other membrane technologies in that the driving force for	The Netherlands	NA	MD	WT/SW	Air gap membrane distillation is a promising desalination process as it is the most efficient MD process. Because a

(continued)

Authors	Year	Title	Journal name, volume (issue)	Abstract	Country	Funding Agent	Technology used	USE (WT/WWT)	Recommendations
				desalination is the difference in vapour pressure of water across the membrane, rather than total pressure.					continuum of evaporative stages in an almost ideal counter-current flow process is occurring in a Memstill® module, a very high recovery of evaporation heat is possible. The Memstill® technology is in operation on pilot plant scale and must still be proven on a larger scale.
Reith, C., Birkenhead, B.	1998	Membranes enabling the affordable and cost effective reuse of wastewater as an alternative water source	Desalination 117 (1998) 203–210	Wastewater reuse is now not only necessary and possible but more surprisingly is affordable and cost effective.	UK	NA	MF	WWT/ Municipal	The reuse of wastewaters is absolutely necessary due to the stress on our most natural of resources from over abstraction, contamination, seawater intrusion and changes in the earth's climate.
Darton, E.G., Buckley, E.	2001	Thirteen years' experiences treating a seawater RO plant	Desalination 134 (2001) 55–62	His paper reviews the long-term operation of the seven 240mVd seawater RO units at Glen Rocky. The paper also considers a few of the unforeseen problems the operator has experienced.	UK	NA	RO	WT/SW	Potable water production from seawater using membranes is a proven technology and a new contract for a 2,000m^3/d plant will be awarded later this year, with the emphasis on reducing produced water costs.
Deane, J.	2002	Private sector participation in desalination in the Mediterranean Middle East (MIME) - past, present and future	Desalination 152 (2002) 57–66	The first part of this paper outlines the principal types of private sector involvement employed on infrastructure projects globally today - from management	UK	NA	RO	WT/SW	The MME countries identified as requiring additional desalination capacity in the foreseeable future are: Algeria, Egypt, Israel, Jordan, Libya and Tunisia.

(*continued*)

Authors	Year	Title	Journal name, volume (issue)	Abstract	Country	Funding Agent	Technology used	USE (WT/WWT)	Recommendations
				contracts to concessions to full-scale privatisations. The second part of the paper sets out the general drivers and constraints behind private sector involvement on desalination projects from both public and private sector perspectives.					
Ragab, R., Prudhomme, C.	2002	Climate Change and Water Resources Management in Arid and Semi-arid Regions: Prospective and Challenges for the 21st Century	Biosystems Engineering (2002) 81(1), 3–34	The overgrowing population and the recent droughts are putting water resources under pressure and calling for new approaches for water planning and management if escalating conflicts are to be avoided and environmental degradation is to be reversed.	UK	FAO, CRU, IPCC-DDC, Prof. Brian Witney, Land Technology Ltd., The support of Dr David Viner at CRU and Dr Nick Reynard. The Environment Agency (EA), London.	desalination	WT&WWT	In a global vision with positive attitude towards water use, sharing and equitability, by year 2050.
Miranda, M.S., Infield, D.	2002	A wind-powered seawater reverse-osmosis system without batteries	Desalination 153 (2002) 9–16	The development of small-scale stand-alone desalination systems is important to communities on islands and in isolated inland areas.	UK	Dulas Ltd, Machynlleth, UK and was supported by ETSU, DTI, UK and CNPq, Department of Science and Technology, Brazil.	RO	WT/SW	Finally, a more representative validation of the system will be only possible after prototype testing, which will follow in the next phase of the project.
Thomson, M., Infield, D.	2002	A photovoltaic-powered seawater reverse-osmosis system without batteries	Desalination 153 (2002) 1–8	An efficient cost-effective battery less photovoltaic-powered seawater reverse-osmosis desalination system is described.	UK	Dulas Ltd, Machynlleth, UK and was supported by ETSU, DTI, UK.	RO	WT/SW	The performance predictions for the system as a whole are based on a detailed Matlab-Simulink model and the critical components of this model

(*continued*)

Authors	Year	Title	Journal name, volume (issue)	Abstract	Country	Funding Agent	Technology used	USE (WT/WWT)	Recommendations
									are based on laboratory measurements. CREST is now commissioning a fully instrumented prototype of the system, in order to verify its performance. Subsequently, field trials will be undertaken.
Howell, J.A.	2004	Future of membranes and membrane reactors in green technologies and for water reuse	Desalination 162 (2004) 1–11	There are many water shortage problems currently in the world. In several of them membrane technology can make a great contribution since membranes have the ability to produce water of exceptional purity that can be recycled for reuse in a variety of places.	UK	NA	NF,MF,RO& chemical treatment.	WWT/ Industrial	Membrane bioreactors will continue to penetrate waste treatment, with the drivers being: *the need for compact plant *simple uprating of capacity *the quality of the effluent * the value of recycling.
Lodge, B., Judd, S.J., Smith, A.J.	2004	Characterisation of dead-end ultrafiltration of biotreated domestic wastewater	Journal of Membrane Science Volume 231, Issues 1–2, 1 March 2004, Pages 91–98	The fouling properties of biologically pretreated greywater and sewage matrices are compared with reference to measured values of specific cake resistance and cake compressibility.	UK	Engineering and Physical Sciences Research Council and Thames Water Utilities Limited.	UF	WWT/ Domestic	It must be concluded that recovery of high-quality membrane-filtered water from foul water recycling is more economically viable than that from greywater recycling.
Dastgir, M.G., Peeva, G., Livingston, A.G.	2005	The performance of composite supported polymeric liquid membranes in the Membrane Aromatic Recovery System (MARS)	Chemical Engineering Science 60 (2005) 7034–7044	This study investigates the possibility of using a composite-supported polymeric liquid membrane (CSPLM) configuration for the	UK	UK Engineering and Physical Sciences Research Council (EPSRC), Muhammad Ghulam Dastgir acknowledges financial support from Ministry of	(CSPLM) & (MARS).	WWT	The long term operational stability under the MARS process conditions implies that this membrane could be scaled up for application in the MARS process. The partition coefficient of

(*continued*)

Authors	Year	Title	Journal name, volume (issue)	Abstract	Country	Funding Agent	Technology used	USE (WT/WWT)	Recommendations
				Membrane Aromatic Recovery System (MARS).		Science and Technology (S&TR Division), Government of Pakistan, The authors also thank GKSS (Germany) for kindly supplying membrane samples.			phenol between PPG/water was measured as 84. A marked effect of ionic strength on partition coefficient of phenol was observed with partition coefficient increase to 134 when measured in the presence of 20 wt% KCl in phenol aqueous solution.
Hankins, N., Hilal, N., Ogunbiyi, O.O., Azzopardi, B.	2005	Inverted polarity micellar enhanced ultrafiltration for the treatment of heavy metal polluted wastewater	Desalination 185 (2005) 185–202	Membrane separation technologies are being applied with increasing frequency in the treatment of industrial and municipal wastewater, as a means of coping with increasingly stringent environmental regulations regarding wastewater discharge.	UK	UK Engineering and Physical Sciences Research Council (EPSRC).	MEUF	WWT/ Municipal& Industrial	The MEUF experiments carried out are proof of those conditions expected to cause 'apparent charge inversion'. In addition, the experiments showed a decrease in fouling relative to those conditions in which charge inversion is absent.
Rachwal, T., Judd, S.	2006	A synopsis of membrane technologies in UK municipal potable water treatment: history, status and prospects	Water and Environment Journal 2006, Print ISSN 1747–6585	The current status of membrane filtration for potable water and wastewater applications is assessed with reference to recent history and key legislative drivers.	UK	UK water suppliers. The Membrane Academia Industry Network.	MF&UF.	WWT/ Municipal	Growth in membrane technologies in the municipal sector is likely to be advanced by a combination of decreasing process costs, increasing stringency of environmental legislation and further process innovation, such as the submerged membrane module.
Lodge, B.N., Judd, S.J.	2006	A statistical approach to the optimisation of membrane operation	Water and Environment Journal 20 96 (2006) 96–100	This paper presents a rigorous analytical technique to identify the process parameters that have	UK	Engineering and Physical Sciences Research Council and Thames Water Plc.	UF	WWT/ Domestic	The results indicate that when the water quality information can be closely related to real-time fouling data, this parameter greatly

(*continued*)

Authors	Year	Title	Journal name, volume (issue)	Abstract	Country	Funding Agent	Technology used	USE (WT/WWT)	Recommendations
				a significant influence on a membrane system performance.					improves the descriptive powers of the analysis.
Rachwal, T., Judd, S.	2006	A synopsis of membrane technologies in UK municipal potable water treatment: history, status and prospects	Water and Environment 20(3) (2006) 110–113	The current status of membrane filtration for potable water and wastewater applications is assessed with reference to recent history and key legislative drivers.	UK	UK MAIN database and originally provided by the UK water suppliers, whose help and advice the authors gratefully acknowledge. The Membrane Academia Industry Network was sponsored by the Engineering and Physical Sciences and Research Council.	UF, MF	WT&WWT	Growth in membrane technologies in the municipal sector is likely to be advanced by a combination of decreasing process costs, increasing stringency of environmental legislation and further process innovation, such as the submerged membrane module.
McAdam, E.J., Judd, S.J.	2006	A review of membrane bioreactor potential for nitrate removal from drinking water	Desalination 196 (2006) 135–148	This paper reviews the research to date appraising advantages and disadvantages associated with each configuration whilst also identifying areas that require further research and consequently assessing which nitrate removal MBR technologies will prevail for the future.	UK	European Physical Sciences and Research Council (EPSRC), Anglian Water, Yorkshire Water and Severn Trent Water.	membrane bioreactor	WT/DW	Greater consideration should be given to the development of submerged pressure-driven MBR, which has proven to be more cost effective and have a lower fouling propensity than the sidestream process in municipal wastewater applications.
Banasiak, L.J., Kruttschnitt, T.W., Schäferc, A.I.	2007	Desalination using electrodialysis as a function of voltage and salt concentration	Desalination 205 (2007) 38–46	In this study the performance of ED for the desalination of different salt solutions and the removal of fluoride and nitrate removal from a brackish groundwater was conducted. The influence of applied	UK	Australian Research Council Linkage collaboration with Brisbane Water as well as the ARC Discovery, Eurodia Germany and France, Berghof and Wytze Meindersma from the University of Twente, Netherlands	Electrodialysis	WT/ Brackish ground-water	Further work is required to determine fouling mechanisms as well as trace contaminant removal mechanisms.

(continued)

Authors	Year	Title	Journal name, volume (issue)	Abstract	Country	Funding Agent	Technology used	USE (WT/WWT)	Recommendations
				voltage and initial salt concentration of a solution was also determined.		and Bart van der Bruggen at the University of Leuven.			
Elimelech, M., Chen, W.H., Waypa, J.J.	1994	Measuring the zeta (electrokinetic) potential of reverse osmosis membranes by a streaming potential analyzer	Desalination, 95 (1994) 269–266	Fouling is a major problem in RO operation, resulting in product water flux decline and larger operating costs. Colloids and particulate matter are thought to cause the most fouling.	USA	State of California, Department of Water Resources (DWR), and by the National Water Research Institute (NWRI). We would like to thank Brian Smith and Kurt Kovac from DWR and Ronald Linski from NWRI for their support and encouragement.	RO	W&WWT	Future research in this area should focus on measuring the zeta potential of pure cellulose acetate and aromatic polyamide polymers, rather than using commercial membranes whose surface properties are not completely known. It is also suggested that the effect of various cleaning techniques on the zeta potential of RO membranes be investigated.
Glater, J., Hong, S.K., Elimelech, M.	1994	The search for a chlorine-resistant reverse osmosis membrane	Desalination, 95 (1994) 325–345	This paper will stimulate research efforts toward development of polymeric reverse osmosis membranes with high levels of chlorine resistance.	USA	State of California, Department of Water Resources (DWR). Our thanks are also extended to Brian Smith of DWR.	RO	WWT/ Industrial	In addition to chlorine sensitivity, future membrane development should focus on a better understanding of the chemical basis for water transport and salt passage. Consideration of both phenomena must proceed hand in hand since they share equal importance in membrane function and durability. There is indeed much work to be done before a truly “ideal” RO membrane becomes a reality.
Childress, A.E., Elimelech, M.	1996	Effect of solution chemistry on the surface charge of	Journal of Membrane Science	The main objective of this paper is to investigate the effect	USA	We acknowledge the support of the U.S. Bureau of Reclamation,	RO&NF	Chemical solution	A at relatively low surfactant concentrations. The striking change in

(continued)

Authors	Year	Title	Journal name, volume (issue)	Abstract	Country	Funding Agent	Technology used	USE (WT/WWT)	Recommendations
		polymeric reverse osmosis and nanofiltration membranes	119 (1996) 253–268	of solution chemical composition on the surface charge properties of various modem commercial RO and NF membranes.		Water Treatment Engineering and Research Group, and of Fluid Systems Corporation, a member of Anglian Water Group. The findings reported in this paper do not necessarily reflect the views of these agencies and no official endorsement should be inferred.			membrane surface charge in the presence of surfactants is attributed to the formation of hemimicelles at the membrane-solution interface.
Hong, S., Elimelech, M.	1997	Chemical and physical aspects of natural organic matter (NOM) fouling of nanofiltration membranes	Journal of Membrane Science 132 (1997) 159–181	The objective of this paper is to investigate systematically the role of chemical and physical interactions in NOM fouling of NF membranes.	USA	We acknowledge the State of California, Department of Water Resources, for financial support and Fluid Systems Corporation, a member of Anglian Water Group, for providing the NF membrane. We also thank Satoshi Tanaka for his help on characterization of NOM and NF membranes.	NF	WT/NOM	Our results demonstrate that cleaning with strong chelating agents, such as EDTA, most effectively removes the fouling layer and restores permeate flux.
Elimelech, M., Zhu, X., Childress, A.E., Hong, S.	1997	Role of membrane surface morphology in colloidal fouling of cellulose acetate and composite aromatic polyamide reverse osmosis membranes	Journal of Membrane Science 127 (1997) 101–109	Fouling of cellulose acetate and aromatic polyamide thinfilm composite reverse osmosis (RO) membranes membranes was studied at identical initial permeation rates so that the effect of the	USA	State of California Department of Water Resources, the National Water Research Institute, U. iversity of California Water Resources Center and Nathan Lewis and Guruswany Kumaraswamy of the Beckman Institute,	RO	WT/Natural	Colloidal fouling of RO membranes is markedly influenced by membrane surface morphology. Surface roughness increases membrane fouling by increasing the rate of colloid attachment onto the membrane surface. The higher fouling rate of aromatic polyamide

(*continued*)

Authors	Year	Title	Journal name, volume (issue)	Abstract	Country	Funding Agent	Technology used	USE (WT/WWT)	Recommendations
				transverse hydrodynamic force (permeation drag) on the fouling of both membranes is comparable.		Material Science Center at Caltech.			composite membranes compared to that of cellulose acetate membranes is attributed to the inherent surface roughness of interfacially-polymerized aromatic polyamide composite membranes.
Zhu, X., Elimelech, M.	1997	Colloidal Fouling of Reverse Osmosis Membranes: Measurements and Fouling Mechanisms	Environ. Sci. Technol. 1997, 31, 3654–3662	The objective of this paper is to systematically investigate the effects of chemical and physical interactions, and the interrelation between such interactions, on the rate of fouling of cellulose acetate and composite polyamide reverse osmosis membranes by silica colloids.	USA	State of California Department of Water Resources, T.e National Water Research Institute, U. iversity of California Water Resources Center, We are also grateful to Nathan Lewis and Guruswany Kumaraswamy of the California Institute of Technology (Beckman Institute, Material Science Center).	RO,NF,UF & MF	W&WWT	because of the inherent permeation drag, fouling is inevitable for source waters with significant levels of colloids. Hence, research should focus on means to effectively clean fouled RO membranes or to optimize pretreatment of source waters.
Tchobano glous, G., Darby, J., Bourgeous, K., McArdle, J., Genest, P., Tylla, M.	1998	Ultrafiltration as an advanced tertiary treatment process for municipal wastewater	Desalination 119 (1998) 315–322	The following paper described Treatment of secondary and tertiary municipal wastewater by membrane filtration has been investigated at the University of California, Davis, California, USA.	USA	NA	UF	WWT/ Municipal	Additional pilot plant testing is planned to further investigate this approach.
Reiss, C.R., Taylor, J.S., Robert, C.	1999	Surface water treatment using nanofiltration – pilot testing results and design considerations	Desalination 125 (1999) 97–112	This research focuses on the factors affecting successful multi-stage pilot-scale operation of IMSs to treat a highly organic surface water.	USA	Tampa Water Department, U.EPA	NF	WT/Surface Water	The ability to restore membrane productivity, T.e benefit of a low fouling CA membrane must be weighed against the potential for damage or destruction via

(*continued*)

Authors	Year	Title	Journal name, volume (issue)	Abstract	Country	Funding Agent	Technology used	USE (WT/WWT)	Recommendations
									biodegradation or hydrolysis. Fouling rates and finished water quality are dependent on membrane make and model, T.e ability of a given membrane to withstand extended exposure to monochloramine.
Vane, L.M., Alvarez, F.R., Giroux, E.L.	1999	Reduction of concentration polarization in pervaporation using vibrating membrane module	Journal of Membrane Science 153(2) (1999) 233–241	A vibrating membrane module currently marketed for ®ltration applications was evaluated for the separation of volatile organic compounds (VOCs) from aqueous solutions by pervaporation.	USA	Laura Yeh of the Naval Facilities Engineering Service Center.	Pervaporation	WT/VOC	VSEP may have to be modified to accept the existing pervaporation membrane materials or new membranes developed to work specifically with a vibrating membrane unit. A field unit fabricated with up to 20m^2 of 250mm thick silicone-coated fiberglass fabric is currently under consideration.
Childress, A.E., Elimelech, M.	2000	Relating Nanofiltration Membrane Performance to Membrane Charge (Electrokinetic) Characteristics	Environ. Sci. Technol. 2000, 34, 3710–3716	The objective of this paper is to systematically investigate the performance (i.e., water flux and solute rejection) of a thin-film composite aromatic polyamide nanofiltration membrane under a wide range of solution chemistries and to relate the performance of the membrane to its chemical and charge characteristics.	USA	the U.S. Bureau of Reclamation, Water Treatment and Engineering Research Group.	NF	W&WWT	The result is substantially decreased water flux and increased salt rejection. This situation would not be as likely to occur at high pH because the membrane is negatively charged, and the hydrocarbon chains will be attracted to the membrane surface leading to adsorption of individual molecules (23). Therefore, only a minimal effect of the surfactant ions is observed at high pH.

(*continued*)

Authors	Year	Title	Journal name, volume (issue)	Abstract	Country	Funding Agent	Technology used	USE (WT/WWT)	Recommendations
Kenneth, H., Carlson, K.H., Amy, G.L.	2000	The importance of soluble microbial products (SMPS) in biological drinking water treatment	Wat. Res. Vol. 34, No. 4, pp. 1386–1396, 2000	The research discussed in this paper had two objectives: 1. determine the magnitude of SMP production relative to the bio-removal of DOC in a typical drinking water biofiltration process; 2. determine the signaficance of SMPs in the DOC pool of the biofilter effluent water.	USA	This work was supported by the Fort Collins (CO) Water Utility and the University of Colorado-Boulder.	Biofilter	WT/DW	SMPs were found to be important only if considered relative to the DOC that is removed during biofiltration. SMPs do not appear to be important if the BOM fraction of the filter-influent DOC is small.
Beckers, H., Diels, L., Stoops, L.	2001	Hollow -fiber technology advances Microbiological Disinfection Capabilities of Low-Pressure MF & UF Membranes	Water Conditioning & Purification, July 2002	Extensive microbiological tests have shown that the combination of an activated carbon prefilter followed by a micro- or ultrafiltration capillary membrane cartridge is an effective barrier against bacteria and (in the case of ultrafiltration) also against viruses.	USA	NA	MF,RO,NF & UF	W&WWT	The combination of an activated carbon prefilter followed by a MF or UF capillary membrane cartridge offers an effective protection against bacteria and (in the case of UF) also meets the USEPA 4-log virus removal requirement established for water purifiers and is in most cases an alternative to other technologies such as RO.
Ergas, S.J., Reuss, A.F.	2001	Hydrogenotrophic denitrification of drinking water using a hollow fibre membrane bioreactor	Journal of Water Supply: Research and Technology—AQUA | 50.3 | 2001	The objective of this research was to investigate the performance of a hollow fibre membrane bioreactor (HFMB) for hydrogenotrophic denitrification of	USA	National Science Foundation.	HFMBR	WT/DW	Periodic shearing of microbial biomass from the surface of the membranes to achieve sustained high denitrification rates is currently under investigation. Hydrogen utilization efficiencies averaged 40%; however, increased H2 utilization

(*continued*)

Authors	Year	Title	Journal name, volume (issue)	Abstract	Country	Funding Agent	Technology used	USE (WT/WWT)	Recommendations
				contaminated drinking water.					efficiency is possible through the use of sealed-end hollow fibre membranes.
Brandhuber, P., Gary Amy, G.	2001	Arsenic removal by a charged ultrafiltration membrane influences of membrane operating conditions and water quality on arsenic rejection	Desalination 140 (2001) 1–14	The purpose of this paper is to investigate the effects of the following conditions on arsenic rejection when the primary mechanism of rejection is Donnan Exclusion: • variation of permeate flux, recovery, cross flow velocity and temperature • variation of arsenic concentration and speciation, • influences of water quality including pH, and the presence of co-occurring, co- and counterions • possible influences of the presence of natural organic matter (NOM).	USA	American Water Works Association Research Foundation.	UF	WT/DW	This work illustrates that relatively 'loose' charged membranes are capable of partial rejection of As(V) from feed waters via Donnan Exclusion.
Vrijenhoek, E.M., Hong, S., Elimelech, M.	2001	Influence of membrane surface properties on initial rate of colloidal fouling of reverse osmosis and nanofiltration membranes	Journal of Membrane Science 188 (2001) 115–128	Recent studies have shown that membrane surface morphology and structure influence permeability, rejection, and colloidal fouling behavior of reverse	USA	American Water Works Association Research Foundation. Dr. Amy Childress of the University of Nevada, Reno for performing the contact angle measurements; Hydranautics,	RO&NF.	WT/SW	AFM images revealed that particles preferentially accumulate in the valleys of rough membranes, leading to 'valley clogging' and causing more severe flux decline than in smooth membranes.

(*continued*)

Authors	Year	Title	Journal name, volume (issue)	Abstract	Country	Funding Agent	Technology used	USE (WT/WWT)	Recommendations
				osmosis (RO) and nanofiltration (NF) membranes.		Dow-FilmTec, Trisep, and Osmonics for supplying membranes; and Nissan Chemical America Corp. for providing the colloidal silica particles.			
Lee, H., Amy, G., Cho, J., Yoon, Y., Moon, S.H., Kim, I.S.	2001	Cleaning strategies for flux recovery of an cleaning ultrafiltration membrane fouled by natural organic matter	Wat. Res. Vol. 35, No. 14, pp. 3301–3308, 2001	One of the most common problems encountered in water treatment applications of membranes is fouling. Natural organic matter (NOM) represents a particularly problematic foulant.	USA	AWWA Research Foundation, Korea Science and Engineering foundation (KOSEF).	UF	WT/NOM	Foulants derived from a hydrophobic NOM source (OC-GW) were cleaned more effectively in terms of flux recovered by acid and caustic cleanings compared to foulants from a relatively hydrophilic NOM source (HT-SW).
Will, M., Klinko, K.	2001	Optimization of seawater RO systems design	Desalination 138 (2001) 299–306	The trend of investment and water cost in RO seawater systems is being evaluated.	USA	NA	RO	WT/SW	Designing seawater RO system for recovery rate exceeding 55% can only provide cost benefits in cases of low feed salinity and low electricity cost.
Decarolis, J., Hong, S., Taylor, J.	2001	Fouling behavior of a pilot scale inside-out hollow fiber UF membrane during dead-end filtration of tertiary wastewater	Journal of Membrane Science Volume 191, Issues 1–2, 30 September 2001, Pages 165–178	A series of pilot-scale filtration experiments were performed systematically under various operating conditions to investigate the fouling behavior of ultrafiltration (UF) membranes to treat tertiary wastewater for resuse.	USA	Hydranautics, D. partment of Molecular Biology and Microbiology at the University of Central Florida, Environmental Systems Engineering Institute (UCF).	UF	WWT/ Municipal	Membrane fouling increased significantly as operating flux increased. The continuous in-line addition of ferric chloride prior to UF filtration significantly reduced the rate of Kw decline for experimental flux values ranging from 54 to 88lmh (32–52gfd).
MacHarg, J.P.	2001	Exchanger Tests Verify 2.0kWh/m^3 SWRO Energy Use	Reprinted from international desalination & water reuse quarterly Vol. 11/1	The PX unit utilises the principle of positive displacement to transfer the energy in the reject stream	USA	NA	RO	WT/Sea water	It has been proved that an ordinary commercial SWRO plant, using a low-recovery, low-pressure design and using Energy

(*continued*)

Authors	Year	Title	Journal name, volume (issue)	Abstract	Country	Funding Agent	Technology used	USE (WT/WWT)	Recommendations
				directly to the feed stream. This direct connection allows a real net energy transfer efficiency from the reject stream to the feed stream of over 95%. The PX device uses a cylindrical rotor with longitudinal ducts parallel to its rotational axis to transfer the pressure energy from the concentrate/reject stream to the feed stream.					Recovery's PX Pressure Exchanger, can operate continuously while achieving energy consumption of 2.0kWh/m^3, and lower.
Mirliss, M.J.	2002	A Comparison of DE and Crossflow Filtration: Just the Facts, Ma'am	Water Conditioning & Purification, July 2002	This article intends to provide an objective comparison of diatomaceous earth (DE) filtration with crossflow (membrane) filtration to allow informed decisions on use of these technologies for municipal, commercial and some industrial applications.	USA	NA	DE filtration.	WWT/ Municipal	DE filtration is simple and inexpensive to evaluate, install and operate. Crossflow filtration, by comparison, has a limited and somewhat stormy history with respect to use in municipal, And when particles below 1μm (sub micron) must be removed, crossflow filtration can be most effective.
Her, N., Amy, G., Foss, D.,Cho, J., Yoon, Y., Kosenka, P.	2002	Optimization of Method for Detecting and Characterizing NOM by HPLC-Size Exclusion Chromatography with UV and On-Line DOC Detection	Environ. Sci. Technol. 2002, 36, 1069–1076	The objective of this research was to evaluate, optimize, and apply a newly modified DOC detector to the HPSECUVA system for characterization of NOM from a wide range of water samples, including treatment process train profiles.	USA	Ionics-Sievers Instruments and technical upport. Philip Brandhuber of the University of Colorados.	HPSEC	WT/NOM	Current research is focused on the application of this method to analyzing samples obtained throughout various water treatment process trains.

(*continued*)

Authors	Year	Title	Journal name, volume (issue)	Abstract	Country	Funding Agent	Technology used	USE (WT/WWT)	Recommendations
Her, N., Amy, G., Foss, D., Cho, J.	2002	Variations of Molecular Weight Estimation by HP-Size Exclusion Chromatography with UVA versus Online DOC Detection	Environ. Sci. Technol. 2002, 36, 3393–3399	The research objectives of this study were to compare and contrast the MW variation between UVA detection with the results obtained by the DOC detector and to suggest an alternative MW representation for NOM comprised of multiple fractional components.	USA	Ionics-Sievers Instruments and and Mr. Paul Kosenka.	HPSEC	WT/NOM	UVA detection should only be used for qualitative analysis, such as comparison of the aromaticity of NOM components as a function of MW.
Nicolaisen, B.	2002	Developments in membrane technology for water treatment	Desalination 153 (2002) 355–360	This paper will describe some technological advance in the area of innovative new membranes and application concepts for spiral wound membrane elements. Spiral wound elements span the four commonly defined membrane technologies, which are microfiltration (0.01-O microns), ultrafiltration (500–100,000Da) nanofiltration (100–500Da), and reverse osmosis (up to 100Da).	USA	NA	MF,UF,NF &RO	W&WWT	Membrane technology has undergone a rapid development as it pertains to most applications and in particular as it pertains to water purification.
Yoon, Y., Amy, G., Cho, J., Her, N., Pellegrino, J.	2002	Transport of perchlorate (C104-)	Desalination 147 (2002) 11–17	Measurements of the rejection of perchlorate anion	USA	American Water Works Association Research Foundation.	UF & NF	Aqueous solutions of perchlorate	ClO, rejection by negatively charged NF and UF membranes was greater

(*continued*)

Authors	Year	Title	Journal name, volume (issue)	Abstract	Country	Funding Agent	Technology used	USE (WT/WWT)	Recommendations
		through NF and UF membranes		(ClO_2-) were performed using nanofiltration (NF) and ultrafiltration (UF) membranes.					than expected based on only steric/size exclusions, due to electrostatic exclusion especially with increasing pH and in the absence of co- and counter ions, which significantly reduce the membrane (negative) surface charge.
Rittmanna, B.E., Stilwell, D., Garside, J.C., Amy, G.L., Spangenberg, C., Kalinsky, A., Akiyoshi, E.	2002	Treatment of a colored groundwater by ozone-biofiltration: pilot studies and modeling interpretation	Water Research 36 (2002) 3387–3397	The goal of the work presented here is to document and understand how the several design parameters interact to control performance of the ozone-biofiltration system in terms of BOM.	USA	NA	ozone-biofiltration	WT	Pilot studies demonstrated that ozonation plus biofiltration removed most color from a colored (50cu) water and also achieved significant removal of DOC.
Lee, K.C., Rittmann, B.E.	2002	Applying a novel autohydro genotrophic hollow-fiber membrane biofilm reactor for denitrification of drinking water	Water Research 36 (2002) 2040–2052	A series of pseudo-steady-state experiments on a novel hollow-fiber membrane biofilm reactor used for denitrification of oligotrophic waters, such as drinking water.	USA	Dr. Charles Gantzer (Membran Corporation, Minneapolis, MN), and Consumers Illinois Company (Kankakee, IL) and Membran Corporation.	Hollow fiber Membrane	WT/DW	The results of this study provide an understanding of how the system performs and key information needed to achieve reliable performance that attain these benefits.
Jarusutthirak, C., Amy, G., Croue, J.-P.	2002	Fouling characteristics of wastewater effluent organic matter (EfOM) isolates on NF and UF membranes	Desalination 145 (2002) 247–255	Wastewater effluent organic matter (EfOM) was isolated into different fractions including colloids, and hydrophobic (HPO) and transphilic (TPI) fractions. The EfOM isolates were characterized by different technique.	USA	Ecole Superieure d'Ingenieurs de Poitiers (ESIP), University of Poitiers, France, and to the Saint Julien 1'Ars.	NF&UF	WWT/ Municipal	This study of fouling characteristics of wastewater effluent organic matter (EfOM) contributes to development of membrane filtration in wastewater reuse applications.

(*continued*)

Authors	Year	Title	Journal name, volume (issue)	Abstract	Country	Funding Agent	Technology used	USE (WT/WWT)	Recommendations
Vane, L.M., Alvarez, F.R.	2002	Full-scale vibrating pervaporation membrane unit: VOC removal from water and surfactant solutions	Journal of Membrane Science 202 (2002) 177–193	A commercial-scale vibrating membrane system was evaluated for the separation of volatile organic compounds (VOCs) from aqueous solutions by pervaporation.	USA	Greg Johnson and Marius Neagoe of New Logic International for VSEP engineering and fabrication services, Anurag Mairal and Hans Wijmans of Membrane Technology and Research for pervaporation membrane and for technical assistance, and Craig Patterson of IT Corporation for logistical and engineering support. The mention of trade names or commercial products does not constitute an endorsement or recommendation for use.	Pervaporation	WT/VOC	Detailed in this paper are the VOC removal characteristics of two full-scale vibrating pervaporation modules for aqueous solutions.
Kimura, K., Amy, G., Drewes, J., Watanabe, Y.	2003	Adsorption of hydrophobic compounds onto NF/RO membranes: an artifact leading to overestimation of rejection	Journal of Membrane Science 221 (2003) 89–101	In this study, three surrogate compounds and three NF/RO membranes were tested under cross-flow conditions.	USA	NA	NF/RO	Hydrophobic compounds	the adsorption of hydrophobic compounds on the membrane occurred significantly, especially when compounds were electrostatically neutral. This adsorption must be taken into account when the rejection of hydrophobic organic compounds by membranes is investigated.
Her, N., Amy, G., McKnight, D., Sohn, J., Yoon, Y.	2003	Characterization of DOM as a function of MW by fluorescence EEM and HPLC-SEC using UVA, DOC,	Water Research 37 (2003) 4295–4303	The objective of this research was to develop a high performance size exclusion chromatography	USA	We gratefully thank Dr. Robin J. Fulton for assisting experiments. This work was partially funded by the Center for Membrane Applied	HPLC-SEC	DOM	the organic matter in a high apparent MW range (over 10,000g/mol) was not the same for the different source waters and was effectively identified by the

(continued)

Authors	Year	Title	Journal name, volume (issue)	Abstract	Country	Funding Agent	Technology used	USE (WT/WWT)	Recommendations
		and fluorescence detection		(HPSEC) system combined with multiple detectors to characterize DOM across a broad range of water samples as a function of apparent MW.		Science and Technology at the University of Colorado.			SEC system with multiple detectors.
Yoon, J., Yoon, Y., Amy, G., Cho, J., Foss, D., Kim, T.H.	2003	Use of surfactant modified ultrafiltration for perchlorate (ClO-4) removal	Water Research 37 (2003) 2001–2012	Determinations of perchlorate anion (ClO-4) transport and rejection were performed using a surfactant modified ultrafiltration (UF) membrane.	USA	The authors wish to acknowledge the American Water Works Association Research Foundation (Ms. Traci Case, project manager) for support of this project.	UF	WT	This study was performed to modify a negatively charged membrane with cationic/anionic surfactants for ClO-4 rejection, and to understand perchlorate transport (rejection) phenomena by changes in steric (size) exclusion and electrostatic exclusion, which can be influenced by cationic surfactant conditioning.
Weroński, P., Walz, J.Y., Elimelech, M.	2003	Effect of depletion interactions on transport of colloidal particles in porous media	Journal of Colloid and Interface Science 262 (2003) 372–383	The influence of depletion interactions on the transport of micrometer-sized, negatively charged polystyrene latex particles through porous media was studied by analysis of particle breakthrough curves as a response to short-pulse particle injections to the inlet of a packed column of glass beads.	USA	U.S. National Science Foundation (Grant CTS-9912098).	porous media	colloidal particles	The results indicate that secondary energy wells significantly affect particle transport behavior through porous media.

(continued)

Authors	Year	Title	Journal name, volume (issue)	Abstract	Country	Funding Agent	Technology used	USE (WT/WWT)	Recommendations
Lee, R.-W., Glater, J., Cohen, Y., Martin, C., Kovac, K., Milobar, M.N., Bartel, D.W.	2003	Low-pressure RO membrane desalination of agricultural drainage water	Desalination 155 (2003) 109–120	In the present work they present laboratory and pilot plant investigations to evaluate the feasibility of membrane desalination of agricultural drainage water in the California San Joaquin Valley.	USA	California Department of Water Resources and by a North American Membrane Society Research Fellowship to Mr. Ron-Wai Lee.	RO	Agricultural drainage water	Work presently underway involves expanded field testing and the evaluation of strategies for reducing membrane fouling via the combination of pretreatment strategies and optimization of membrane system configuration, all specifically targeted for desalination of agricultural drainage water in the California San Joaquin Valley.
Sohn, J., Amy, G., Cho, J., Lee, Y., Yoon, Y.	2004	Disinfectant decay and disinfection by-products formation model development: chlorination and ozonation by-products	Water Research 38 (2004) 2461–2478	The purpose of this paper was two fold. First, a comprehensive model was developed for predicting chlorine and ozone residual and DBP formation including THMs, HAAs, and bromate for raw and coagulated waters. Second, this research has provided D/DBP models for disinfectant/DBP simulation programs, such as the U.S. EPA water treatment plant model (WTPM) and the American Water Works Association Research Foundation (AwwaRF) integrated disinfection design framework (IDDF).	USA	The authors wish to acknowledge Zaid Chowdhury and Kenan Ozekin for providing valuable data for this study, and to Generale des Eaux for partial support of this work.	Desalination	WT	In this study, several empirical models for predicting disinfectant (in this study chlorine and ozone) decay and disinfection by-product formation were evaluated and/ or developed.
Lee, N., Amy, G., Croue, J.P., Buisson, H.	2004	Identification and understanding of fouling in low-pressure membrane	Water Research 38 (2004) 4511–4523	Four different membranes were studied in (constant pressure) flux decline	USA	NA	UF/MF/HPO/HPI	WT/NOM	The results indicate that MF membranes are more prone to fouling than UF membranes. Trends

(*continued*)

Authors	Year	Title	Journal name, volume (issue)	Abstract	Country	Funding Agent	Technology used	USE (WT/WWT)	Recommendations
		(MF/UF) filtration by natural organic matter (NOM)		tests for the study of NOM fouling two MF and two UF membranes possessing hydrophilic (HPI) or hydrophobic (HPO) properties, based on contact angle measurements.					according to membrane hydrophobicity/ hydrophilicity are less clear, likely because surface roughness may be a more influential factor than HPO/ HPI character.
Her, N., Amy, G., Park, H.R., Song, M.	2004	Characterizing algogenic organic matter (AOM) and evaluating associated NF membrane fouling	Water Research 38 (2004) 1427–1438	The aim of this study were to characterize AOM by various methods and to evaluate NF membrane fouling associated with AOM.	USA	This work was funded by the US National Science Foundation (NSF) Center for Membrane Applied Science and Technology at the University of Colorado, and through partial support of Generale des Eaux (Veolia Water).	NF	WT/AOM	AOM extracted in water as a solvent showed a high HPI fraction (57.3%) and a low SUVA (1.0L/m-mg). Significant amounts of proteinaceous components such as MAAs (at 329 nm by UV/visible absorption) and phycobilin were detected by UV/visible absorption.
Bellona, C., Drewes, J.E., Xu, P., Amy, G.	2004	Factors affecting the rejection of organic solutes during NF/RO treatment—a literature review	Water Research 38 (2004) 2795–2809	In this paper, findings of a comprehensive literature review are reported, targeting membrane rejection mechanisms and factors affecting rejection.	USA	WateReuse Foundation (WRF) for its financial, technical, andad ministrative assistance in funding and managing the project through which this information was derived. The authors also thank the West Basin Municipal Water District for its financial andtechnical support.	NF/RO	WT	Ongoing research efforts are directed to verify the usefulness of this diagram in determining key parameters affecting rejection andalso in quantifying rejection of a certain compoundby a particular membrane type.
Kinslow, J.K., Hudkins, J.M.	2004	The Evolution of Pretreatment Chemicals in	Florida water resources journal • Nov. 2004 • 21–28	A treatment facility can be built and operated using the best practiced	USA	NA	RO	WT/BW	With the advent of new pretreatment chemicals and the continuation of efforts to optimize their use,

(*continued*)

Authors	Year	Title	Journal name, volume (issue)	Abstract	Country	Funding Agent	Technology used	USE (WT/WWT)	Recommendations
		Membrane Processes		methods in the industry and still have operational difficulties if the pretreatment chemicals are not chosen and applied correctly for the conditions of service.					the future holds much promise for new and more challenging applications to come.
Duranceau, S.J., Henthorne, L.R.	2004	Membrane Pretreatment for Seawater Reverse-Osmosis Desalination	Florida water resources journal • Nov. 2004 • 33–36	One of the most significant factors in successfully and costeffectively operating a reverse-osmosis (RO) seawater desalting plant is the ability of the pretreatment system to consistently produce high-quality feedwater and relatively microbe-free water for RO processing.	USA	The project was funded by the U.S. Bureau of Reclamation.	RO, UF, NF, MF	WT/SW	RO seawater desalting performance differed significantly between conventional and membrane treatment trains.
Li, Q., Elimelech, M.	2004	Organic Fouling and Chemical Cleaning of Nanofiltration Membranes: Measurements and Mechanisms	Environ. Sci. Technol. 2004, 38, 4683–4693	In this study, atomic force microscopy was used to quantify the interaction forces between the foulant and the clean or fouled membrane surface.	USA	U.S. Bureau of Reclamation under the Desalination and Water Purification Research and Development Program (DWPR).	NF	WT	The solubilization force is then strong enough to break up all the Ca^{2+} induced bridges leading to release of the humic acid to the aqueous phase. The local micelles formed also help keeping the humic acid molecules from reprecipitating.
Saad, M.A.	2004	Early discovery of RO membrane fouling and real-time monitoring of plant performance for	Desalination 165 (2004) 183–191	This paper discusses the shortfalls of this technique, and introduces a practical new technology for	USA	NA	RO	WT/SW	The new Silent Alarm TM technology can provide these plants and many others a unique tool to help them achieve these

(*continued*)

Authors	Year	Title	Journal name, volume (issue)	Abstract	Country	Funding Agent	Technology used	USE (WT/WWT)	Recommendations
		optimizing cost of water		measuring and monitoring membrane fouling and flux performance in real-time. The new Silent Alarm TM technology, designed as an early-warning system, allows the discovery of any fouling or scaling development on the RO membranes in the very early stages, thus providing a valuable tool for the plant operator to take immediate corrective measures before it is too late.					objectives on a daily basis. More important, this proven capability to monitor RO, nanofiltration and other membrane-based plant performance characteristics in real-time, and discovering, not trending, the actual development of any membrane fouling or scaling before too late has a direct and dramatic positive impact on optimizing the total cost of desalinated water.
Merlo, R.P., Trussell, R.S., Hermanowicz, S. W., Jenkins, D.	2004	Physical, Chemical and Biological properties of submerged membrane bioreactor and conventional activated sludges	hermanowicz@ce.berkeley.edu.	A pilot-scale submerged membrane bioreactor (SMBR) and two bench-scale conventional activated sludge (CAS) reactors were operated on municipal primary effluent over a range of mean cell residence times (MCRTs) from 2–10d.	USA	Water Environment Research Foundation (WERF).	SMBR	WWT/ Municipal	SMBR effluent total COD concentrations were much lower than CAS effluent total COD concentrations at all MCRTs tested because of the absence of TSS in the SMBR effluent and because the SMBR effluent contained much lower soluble CODs than the CAS effluents, possibly because of the retention of some soluble and colloidal organic material by the membrane. At a 5-d MCRT, the SMBR and the high-shear CAS sludges had similar PSDs with high numbers of small particles ($<4\mu m$), The colloidal material

(*continued*)

Authors	Year	Title	Journal name, volume (issue)	Abstract	Country	Funding Agent	Technology used	USE (WT/WWT)	Recommendations
									(approximately $<10\ \mu m$) content of the SMBR sludges was about one order of magnitude greater than in the CAS sludges, The sludge total EPS content and its protein and carbohydrate constituents were all higher in the CAS sludges than in the SMBR. There did not appear to be any difference in normalized TTF for any of the sludges tested.
MacHarg, J.P., McClellan, S.A.	2004	Pressure Exchanger Helps Reduce Energy Costs in Brackish Water RO System	JOURNAL AWWA | NOVEMBER 2004, 44–47	A device from Energy Recovery Inc. (ERI), which has been proven effective in hundreds of seawater reverse osmosis (RO) applications, has now been shown to save money in brackish RO plants as well.	USA	NA	RO-PX	WT/BW	With savings typically ranging between 10 and 30% in lower-pressure, higher-recovery systems, the PX may become as common in the brackish RO market as it is in seawater markets.
Wolf, P.H., Siverns, S.	2004	The New Generation for Reliable RO Pretreatment	International conference on desalination costing, Dec, 2004, Cyprus	This paper discusses the advantages and performance of UF membrane pretreatment and compares overall life-cycle costs.	USA	NA	UF, RO	WT/ Sea water	UF pretreatment systems have proven to be simple to operate, much more reliable and more economical over the life of a project than conventional pretreatment, particularly when the raw feedwater is highly loaded with suspended solids and colloids permanently, or during a part of a season.
Shih, W.-Y., Albrecht, K., Yoram Cohen, Y.	2004	A dual-probe approach for evaluation of	Desalination 169 (2004) 213–221	An experimental system was developed to assess	USA	California Department of Water Resources, M. tropolitan Water District	RO	Agricultural drainage water	The present approach of evaluating AS retardation of the crystallization

(*continued*)

Authors	Year	Title	Journal name, volume (issue)	Abstract	Country	Funding Agent	Technology used	USE (WT/WWT)	Recommendations
		gypsum crystallization in response to antiscalant treatment		and rank the performance of commercial antiscalants designed to inhibit mineral scale formation on reverse osmosis membrane surfaces.		of Southern California, the California Energy Commission and by an International Desalination Association (IDA) Research Fellowship to Ms. Wen-Yi Shih.			induction time should allow rapid screening of antiscalants for membrane desalination applications.
Lee, N., Amy, G., Lozi, J.	2005	Understanding natural organic matter fouling in low-pressure membrane filtration	Desalination 178 (2005) 85–93	Natural organic matter (NOM) in natural water is responsible for significant organic fouling of low pressure membranes. Membrane fouling can be affected by the characteristics of the NOM and its interaction with the membranes and their associated properties.	USA	NA	UF & MF	WT/NOM	The waters containing high colloids and macromolecules as well as high DOC content are very problematical in low-pressure membrane fouling.
Bellona, C., Drewes, J.E.	2005	The role of membrane surface charge and solute physico-chemical properties in the rejection of organic acids by NF membranes	Journal of Membrane Science 249 (2005) 227–234	The objective of this study was to investigate the contribution of electrostatic interactions to the rejection of selected organic acids by nanofiltration membranes.	USA	WateReuse Foundation (WRF), Central/West Basin Municipal Water District, Dow/Filmtec for providing membrane specimens, Dr. Jaewon Cho, K-JIST (South Korea), Dr. Gary Amy and Tae-Uk Kim, University of Colorado as well as Dr. Pei Xu, Colorado School of Mines.	NF	Organic acid	The rejection of negatively charged organic acids by NF membranes resulted in a larger rejection than expected based on steric/size exclusions due to electrostatic repulsion between solute and membrane as driving factor for rejection.
Lee, N., Amy, G., Croue, J.P., Buisson, H.	2005	Morphological analyses of natural organic matter (NOM) fouling of low-pressure	Journal of Membrane Science 261 (2005) 7–16	The overall objective of this research is to investigate natural organic matter (NOM) fouling in low-pressure	USA	Veolia water	MF&UF	WT/NOM	Morphological analysis also provides better understanding of two major fouling mechanisms in UF and MF membranes:

(*continued*)

Authors	Year	Title	Journal name, volume (issue)	Abstract	Country	Funding Agent	Technology used	USE (WT/WWT)	Recommendations
		membranes (MF/UF)		membranes (MF/UF) by morphological analyses, with comparison before and after membrane filtration.					surface coverage versus pore blockage, respectively. Clean UF membranes are relatively smother than clean MF membranes, imparting less fouling potential.
Zhao, Y., Taylor, J., Hong, S.	2005	Combined influence of membrane surface properties and feed water qualities on RO/NF mass transfer, a pilot study	Water Research 39 (2005) 1233–1244	The impact of membrane surface characteristics and NOM on membrane performance has been investigated for varying pretreatment and membranes in a field study. Surface charge, hydrophobicity and roughness varied significantly among the four membranes used in the study.	USA	This project was supported by the St. Johns River Water Management District and contracted directly to CH2M Hill, who subcontracted analytical, interpretation and modeling work to UCF. The authors acknowledge SJRWMD and CH2M Hill.	RO/NF	WT	All membranes had uniquely different surface characteristics. The membrane systems were operated in a similar manner and provided a basis for comparing membrane performance as affected by surface characteristics and water quality. Water and inorganic mass transfer increased with contact angle (hydrophobicity) and surface roughness. The initial inorganic Ks's were different for each of the four membranes for similar solutes, Loss of sodium and chloride mass transfer during operation increased as contact angle increased (hydrophobicity) and decreased as roughness decreased.
Elarde, J.R., Weightman, R.B., Beddow, W.D., Bergman, R.A.	2005	Conversion from Membrane Softening to Brackish-Water Reverse Osmosis	Florida water resources journal. Nov. 2005, 51–56	In the spring of 2002, the city of Fort Myers was proactive in developing a new brackish-water supply to replace its existing limited shallow aquifer water source, which also was causing	USA	NA	RO	WT/BW	Converting the Fort Myers membrane softening WTP to a brackish-water RO WTP was a complex process that required changes to several of the WTP unit processes.

(*continued*)

Authors	Year	Title	Journal name, volume (issue)	Abstract	Country	Funding Agent	Technology used	USE (WT/WWT)	Recommendations
				some membrane fouling issues at the existing facility. The city also converted its existing nanofiltration (NF) membrane softening facility to a brackish reverse-osmosis (RO) system and expanded the plant capacity.					
Stover, R.L.	2005	Energy recovery device pereformance analysis	Water Middle East 2005 1 November 14, 2005,1–3	The effectiveness of an energy recovery device can be characterized by quantifying the high-pressure pump energy required to produce a given flow rate of desalinated permeate with an SWRO system equipped with the device.	USA	NA	RO	WT/Sea water	SWRO systems equipped with PX Pressure Exchanger energy recovery technology require very low energy per unit of permeate produced. In addition, the design of the PX device provides the SWRO operator many advantages over other energy recovery devices including no periodic maintenance, consistent performance, and ease of control and operation.
Stover, R.L., Ameglio, A., Khan, P.A.K.	2005	The ghalilah SWRO plant an overview of the solutions adopted to minimize energy consumption	Desalination 000 (2005) 1–5	The plant is the largest SWRO installation in the Arabian Gulf producing three MIGD (13,650 cubic meters) of potable water per day.	USA	NA	RO	WT/ Sea water	The Ghalilah plant is running at its design production rate and producing water at less than 4 kWh/m^3 The PXs are running with low mixing, low noise and high efficiency as predicted during the design stage. PX efficiency exceeds 95% over the operating range of the plant.
Choi, H., Zhang, K., Dionysiou, D. D., Oerther, D.B., Sorial, G.A.	2005	Effect of activated sludge properties and membrane operation conditions	Chemosphere 63(10) (2006) 1699–1708	This study examined the impact of biological, chemical, and physical properties	USA	Ohio State University Research.	MBR	Activated slugde	Results on membrane filtration performance of the activated sludges suggest that membrane biofouling is

(*continued*)

Authors	Year	Title	Journal name, volume (issue)	Abstract	Country	Funding Agent	Technology used	USE (WT/WWT)	Recommendations
		on fouling characteristics in membrane bioreactors		of activated sludge on membrane filtration performance in laboratory-scale MBRs.					a complex phenomenon and depends on the physical, chemical, and biological aspects of the activated sludge.
Kim, S., Hoek, E.M.V.	2005	Modeling concentration polarization in reverse osmosis processes	Desalination 186 (2005) 111–128	The objective of this study was to compare available analytical CP models to a more rigorous numerical CP model and experimental CP data.	USA	Korea Science and Engineering Foundation (KOSEF).	RO	Pure water	The use of film theory based concentration polarization models in real RO processes appears justified given the specific operating conditions of high crossflow and low flux used in most practical applications.
Shih, W.-Y., Rahardianto, A., Lee, R.-W., Cohen, Y.	2005	Morphometric characterization of calcium sulfate dihydrate (gypsum) scale on reverse osmosis membranes	Journal of Membrane Science 252 (2005) 253–263	The axial development of calcium sulfate dihydrate (gypsum) scaling on selected reverse osmosis (RO) membrane surfaces was investigated.	USA	California Department of Water Resources, the California Energy Commission, the Metropolitan Water District of Southern California, and by an International Desalination Association (IDA).	RO	WT/ Calcium sulfate dihydrate	The results of the present study suggest that both membrane gypsum scaling and antiscalant action appear to directly affect surface crystallization. Clearly, results of the present study suggest the need for further studies on the impact of surface topology and chemistry on chemistry on surface crystallization.
Li, Q., Elimelech, M.	2006	Synergistic effects in combined fouling of a loose nanofiltration membrane by colloidal materials and natural organic matter	Journal of Membrane Science 278 (2006) 72–82	The study presented here aims to unravel the mechanisms of combined organic/ colloidal fouling of NF membranes.	USA	the National Science Foundation and by the US Bureau of Reclamation under the Desalination and Water Purification Research and Development Program (DWPR).	NF	WT/Natural organic matter (NOM)	Combined fouling of a 'loose', low salt rejection NF membrane by colloidal materials and dissolved NOM was investigated in both dead-end and cross-flow filtration systems.
Boyer, T.H., Singer, P.C.	2006	A pilot-scale evaluation of magnetic ion exchange treatment for removal of	Water Research 40 (2006) 2865–2876	The objective of this research was to evaluate a magnetic ion exchange process (MIEX) for the	USA	Brown Water Treatment Plant in Durham, University of Colorado, Orica Watercare.	Anion exchange	WT/NOM	Substantial removal of NOM was achieved at ERDs as low as 0.16mL/L, Bromide removal of 20–28% was achieved,

(*continued*)

Authors	Year	Title	Journal name, volume (issue)	Abstract	Country	Funding Agent	Technology used	USE (WT/WWT)	Recommendations
		natural organic material and inorganic anions		removal of natural organic material (NOM) and bromide on a continuous-flow pilotscale basis under different operating conditions and raw water characteristics.					Increasing the sulfate concentration in the raw water to approximately 50 mg/L resulted in decreased removal of DOC, The fraction of resin saturation increases with decreasing ERD.
Lee, S., Ang, W.S., Elimelech, M.	2006	Fouling of reverse osmosis membranes by hydrophilic organic matter: implications for water reuse	Desalination 187 (2006) 313–321	Effluent organic matter (EfOM) is suspected as a major cause of fouling of reverse osmosis (RO) membranes in advanced wastewater reclamation.	USA	US Department of Interior, Bureau of Reclamation, Desalination Research and Development Program, Post-doctoral Fellowship Program of Korea Science and Engineering Foundation (KOSEF).	RO	WWT/ Municipal	Increasing initial flux results in higher rate of alginate fouling, whereas increasing crossflow velocity lessens the extent of fouling. Thus, optimization of hydrodynamic conditions can be useful for organic fouling minimization.
Rittmann, B.E.	2006	The membrane biofilm reactor: the natural partnership of membranes and biofilm	The membrane biofilm reactor: the natural partnership of membranes and biofilm.	The MBfR has been demonstrated for denitrification in drinking water; reduction of perchlorate in groundwater; reduction of selenate, chromate, trichloroethene and other emerging contaminants; advanced N removal in wastewater treatment and autotrophic total-N removal.	USA	US Environmental Protection Agency, American Water Works Association Research Foundation, Applied Process Technology Inc., National Water Research Institute, Inter-State Water Co., Membran Corp., Mitsubishi-Rayon Corp., Montgomery-Watson-Harza and Metcalf & Eddy of New York.	MBfR	WWT/ Municipal	Gas pressures must be controlled properly, but proper control makes it possible to produce an effluent with total inorganic N of about 1 mg/L with a totally autotrophic MBfR system.
Harper (Jr), WF., Bernhardt, M., Newfield, C.	2006	Membrane bioreactor biomass characteristics and microbial yield at	Water SA Vol. 32 No. 2 April 2006	This study confirmed that the MBR maintains higher solids levels and	USA	Auburn University Environmental Institute and Auburn University	MBR	WWT/ Municipal	The MBR true yield was higher than that of the SBR; future experiments should focus on electron

(*continued*)

Authors	Year	Title	Journal name, volume (issue)	Abstract	Country	Funding Agent	Technology used	USE (WT/WWT)	Recommendations
		very low mean cell residence time		better overall effluent quality than conventional bioreactors at all MCRTs tested.		undergraduate research program.			transfer efficiency as the underlying reason for this observation. This study expands the MBR dataset available for very low MCRT operation.
Trussell, R.S., Merlo, R.P., Hermanowicz, S.W., Jenkins, D.	2006	The effect of organic loading on process performance and membrane fouling in a submerged membrane bioreactor treating municipal wastewater	Water Research 40 (2006) 2675–2683	The results of experiments on municipal wastewater primary effluent are presented for a pilot-scale submerged membrane bioreactor (SMBR).	USA	Water Environment Research Foundation (WERF).	MBR	WWT/ Municipal	The membrane fouling rates did not correlate well with soluble COD measured on a 0.45mm membrane filtrate of mixed liquor or with soluble COD rejection (effluent COD/ soluble COD).
Subramani, A., Kim, S., Hoek, E.M.V.	2006	Pressure, flow, and concentration profiles in open and spacer-filled membrane channels	Journal of Membrane Science 277 (2006) 7–17	A finite element model was developed to study momentum and mass transport in cross flow membrane filtration systems with open and spacerfilled channels.	USA	United States Environmental Protection Agency (US EPA) through the Desalination Research Innovation Partnership (DRIP).	RO	WT	On average, feed channel spacers reduced the extent of concentration polarization because of the enhanced wall shear rates. However, the stagnant regions (in front and behind some spacer filaments) led to enhanced concentration polarization, which suggests certain spacer designs could promote local scale and cake formation in spiral wound elements.
Shih, W.-Y., Gao, J., Rahardianto, A., Glater, J., Cohen, Y., Gabelich, C.J.	2006	Ranking of antiscalant performance for gypsum scale suppression in the presence of residual aluminum	Desalination 196 (2006) 280–292	An approach to ranking antiscalant effectiveness for gypsum scale suppression using combined bulk crystallization and membrane scaling	USA	Metropolitan Water District of Southern California and the California Department of Water Resources.	RO	WT/ (Aluminum, Gypsum)	It is suggested that the above approach can be valuable in identifying effective candidate antiscalants, thereby assisting in the design and optimization of field pilot membrane studies.

(continued)

Authors	Year	Title	Journal name, volume (issue)	Abstract	Country	Funding Agent	Technology used	USE (WT/WWT)	Recommendations
				diagnostic tests was explored with a focus on the adverse impact of residual aluminum in dissolved (Al3+) and as colloidal species on gypsum scale suppression.					
Norberg, D., Hong, S., Taylor, J., Zhao, Y.	2007	Surface characterization and performance evaluation of commercial fouling resistant low-pressure RO membranes	Desalination 202 (2007) 45–52	This paper describes the characterization and evaluation of various RO/NF membranes for the treatment of seasonally brackish surface water with high organic contents (TOC ≈21mg/L).	USA	This project was sponsored by the St. Johns River Water Management District (Jerry Salsano: SJRWMD project officer).	RO	WT/Brackish surface water	Chemical compatibility should be carefully considered in the process of developing fouling resistant membranes.
Liu, C.C.K., Xia, W., Park, J.W.	2007	A wind-driven reverse osmosis system for aquaculture wastewater reuse and nutrient recovery	Desalination 202 (2007) 24–30	A prototype wind-driven reverse osmosis (RO) system was constructed in 1997–1998 by University of Hawaii researchers.	USA	U.S.National Science Foundation and by the U.S. Geological Survey.	RO	WWT/ Aquaculture	The wind-driven RO process was applied successfully for nitrogen removal from aquacultural wastewater. It was demonstrated that duckweed plays an important role in the N-absorption process under conditions of both high and low nutrient levels.
Kang, S., Lee, W., Chae, S., Shin, H.	2007	Positive roles of biofilm during the operation of membrane bioreactor for water reuse	Desalination 202 (2007) 129–134	In this study, positive roles of membrane biofilm in the intermittently aerated MBR system were highlighted.	USA	G-7 Project from Ministry of Environment.	MBR	WWT/ Municipal	This study will give unique insight into positive roles of membrane biofilm, and also continue to aid both fundamental studies and developments of MBR processes.
Snyder, S.A., Adham, S.,	2007	Role of membranes and activated	Desalination 202 (2007) 156–181	The primary objective of this investigation	USA	Water Quality Research and	RO	WWT/ Industrial	Animal studies are required to determine the

(*continued*)

Authors	Year	Title	Journal name, volume (issue)	Abstract	Country	Funding Agent	Technology used	USE (WT/WWT)	Recommendations
Redding, A.M., Cannon, F.S., DeCarolis, J., Oppenheimer, J., Wert, E.C., Yoon, Y.		carbon in the removal of endocrine disruptors and pharmaceuticals		was to determine the efficacy of various membranes and activated carbons for the removal of endocrine disruptors, pharmaceuticals, and personal care products.		Development Division of the Southern Nevada Water Authority.			toxicologically relevant levels of these emerging contaminants and to establish meaningful treatment goals.
Uchymiak, M., Rahardianto, A., Lyster, E., Glater, J., Cohen, Y.	2007	A novel RO ex situ scale observation detector (EXSOD) for mineral scale characterization and early detection	Journal of Membrane Science 291 (2007) 86–95	A novel RO ex situ scale observation detector (EXSOD) was developed for direct visual real-time monitoring of mineral surface scaling on RO membranes.	USA	California Department ofWater Resources, the NationalWater Research Institute, the USEPA, the University of California Center for Water Resources, the Desalination Research Innovation Partnership and by graduate student scholarships awarded to E. Lyster by the National Water Research Institute and to A. Rahardianto by the International Desalination Association.	RO	WT	It is anticipated that on-line monitoring suggested in the present study will prove useful in establishing the frequency of cleaning regimens and thus extend the useful life of membrane modules.
Rahardianto, A., Gao, J., Gabelich, C.J., Williams, M.D., Cohen, Y.	2007	High recovery membrane desalting of low-salinity brackish water: Integration of accelerated precipitation softening with membrane RO	Journal of Membrane Science 289 (2007) 123–137	The integration of membrane ROdesalting and accelerated precipitation softening (APS)was evaluated for achieving high productwater recovery (>95%) from desalting of mildly brackish surface water.	USA	California Department of Water Resources, the Metropolitan Water District of Southern California, the California Energy Commission, the California Water Resources Center, and the U.S. Environmental Protection Agency.	RO	WT/BW	Diagnostic membrane scaling tests with the final secondary RO concentrate demonstrated that APS treatment is technically feasible for high recovery CR water desalting.

IWA Publishing's authorised EU representative for General Product Safety Regulations is Diane D'Arras, 15 rue Duret, 75116 Paris, France, e-mail: safety@iwap.co.uk.

Printed and bound by CPI Group (UK) Ltd, Croydon, CR0 4YY
12/05/2026
02108876-0018